ADHESION AND BONDING

ADHESION AND BONDING

Encyclopedia Reprints

Edited by

NORBERT M. BIKALES

Consulting Chemist
Livingston, New Jersey

WILEY-INTERSCIENCE

a Division of John Wiley & Sons, Inc.
New York · London · Sydney · Toronto

Library of Congress Catalog Card Number: 78-172950

ISBN 0-471-07230-3

Printed in the United States of America.

10 9 8 7 6 5 4 3 2 1

CONTRIBUTORS

JEROME L. BEEN, Talon Adhesives Corporation, Kearny, New Jersey, *Bonding*

NORBERT M. BIKALES, Consulting Chemist, Livingston, New Jersey, *Editor*

J. J. BIKERMAN, 15810 Van Aken Boulevard, Cleveland, Ohio (formerly at Massachusetts Institute of Technology), *Theory of Adhesive Joints*

R. F. BLOMQUIST, Forest Sciences Laboratory, U.S. Department of Agriculture, Athens, Georgia, *Applications; Evaluation*

EINO MOKS, Weyerhaeuser Company, Seattle, Washington, *Analysis of Adhesives*

GEORGE P. KOVACH, Koro Corporation, Hudson, Massachusetts (formerly at Foster Grant Co.), *Abherents*

MILTON ROTHSTEIN, The Thermatron Company, Willcox & Gibbs Sewing Machine Co., Bay Shore, New York, *Dielectric Heating*

HAROLD SCHONHORN, Bell Telephone Laboratories, Inc., Murray Hill, New Jersey, *Surface Properties*

JEFFREY R. SHERRY, Branson Sonic Power Company, Danbury, Connecticut, *Ultrasonic Fabrication*

IRVING SKEIST, Skeist Laboratories, Inc., Livingston, New Jersey, *Adhesives Compositions*

HARLAND H. YOUNG, Swift & Company, Chicago, Illinois, *Glue, Animal and Fish*

W. A. ZISMAN, U.S. Naval Research Laboratory, Washington, D.C., *Effect of Chemical Constitution*

PREFACE

Bonding of materials, particularly plastics, has been largely an empirical art because the necessary understanding of the phenomenon of adhesion has been lacking. A better theoretical basis has recently begun to be built, although there are still many controversial issues, as can be seen from the different approaches in the first three chapters. These fundamental advances are placing the technology on a more rational foundation.

This book assembles in one volume thirteen chapters from seven different sources dealing with the theory of adhesion and, especially, with practical methods of bonding. The subjects covered are the composition, analysis, and applications of adhesives, the design and evaluation of adhesive joints, methods of welding, and abherents, or release agents. The contributors are recognized authorities in the field.

Because the chapters are reproduced unchanged from their original sources, the reader should ignore the cross references to other pages and to other chapters that appear in the text.

I should like to acknowledge the great help of Jo Conrad in editing the original Encyclopedia articles on which this book is based.

NORBERT M. BIKALES

Livingston, New Jersey
June 1971

CONTENTS

ADHESION AND BONDING

ADHESION AND BONDING

Adhesion (adherence) is the phenomenon in which surfaces are held together by interfacial forces. Adhesion may be *mechanical, electrostatic,* or by *molecular attraction,* depending upon whether it results from interlocking action, from the attraction of electrical charges, or from valence forces, respectively.

This article deals primarily with adhesion by molecular attraction, and more particularly with the theory of *adhesive joints,* or *bonds,* and with the practical aspects of bond formation (*bonding,* sometimes called *gluing* or *cementing*). An important aspect of this subject is the chemical nature of the *adhesive,* ie, the substance capable of holding materials together by surface attachment, and of the *adherends,* ie, the bodies being held together by the adhesive. Both the adhesive and the adherends may be high-molecular-weight materials.

The article begins with two sections that provide the theoretical basis for the subsequent discussion of the applied areas. A description of polymers used in adhesive compositions follows, after which the engineering aspects of bonding are discussed in detail. The article concludes with a survey of applications and of test methods.

GLOSSARY

The following list of terms is based on ASTM D 907-55. Courtesy American Society for Testing and Materials.

Adherend. A body that is held to another body by an adhesive.

Adhesion (adherence). The phenomenon in which surfaces are held together by interfacial forces.

Adhesion, specific. Adhesion between surfaces that are held together by valence forces of the same type as those that give rise to cohesion.

Adhesive. A substance capable of holding materials together by surface attachment. Adhesive is the general term and includes cement, glue, mucilage, paste, etc. These terms are all used essentially interchangeably. Various descriptive adjectives are often applied to the term adhesive to indicate certain characteristics, as for example: physical form, eg, liquid adhesive, tape adhesive; chemical form, eg, silicate adhesive, epoxy adhesive; materials bonded, eg, paper adhesive, metal–plastic adhesive, can-label adhesive; conditions of use, eg, cold-setting adhesive, hot-setting adhesive.

Adhesive, cold-setting. An adhesive that sets at temperatures below 20 °C.

Adhesive, hot-setting. An adhesive that requires a temperature at or above 100 °C to set.

Adhesive, intermediate-temperature-setting. An adhesive that sets in the temperature range 31–99 °C.

Adhesive, pressure-sensitive. An adhesive made to adhere to a surface at room temperature by briefly applied pressure alone.

Adhesive, room-temperature-setting. An adhesive that sets in the temperature range 20–30 °C in accordance with the limits for Standard Room Temperature specified in the Standard Methods of Conditioning Plastics and Electrical Insulating Materials for Testing (ASTM D 618).

Adhesive, solvent-release. An adhesive that releases solvent vapor during cure.

Adhesive, warm-setting. A term sometimes used synonymously with *Intermediate-temperature-setting adhesive.*

Assembly. A group of materials or parts, including adhesive, that have been placed together for bonding or that have been bonded together.

Assembly time. The time interval (either necessary or permissible) between spreading the adhesive to the adherends and the application of pressure or heat, or both, to the assembly. The *open assembly time* is the optimum time interval after applying the adhesive when surfaces should be joined.

Binder. A component of an adhesive composition that is primarily responsible for the adhesive forces that hold two bodies together.

Blister. An elevation of the surface of an adherend, somewhat resembling in shape a blister on the human skin; its boundaries may be indefinitely outlined and it may have burst and become flattened.

Blocking. Undesired adhesion between touching layers of a material, such as occurs under moderate pressure during storage or use.

Bond, v. To attach materials together by an adhesive. n. The attachment at an interface between an adhesive and an adherend. See also *Joint.*

Cement, n. See *Adhesive.* v. See *Bond.* See also *Solvent cement.*

Cleavage. A method of rupturing adhesive bonds between rigid materials, best described as a prying action.

Cohesion. The phenomenon in which the particles of a single substance are held together by primary or secondary valence forces.

Cold pressing. A bonding operation in which an assembly is subjected to pressure without the application of heat.

Conditioning time. The time interval between the removal of the joint from the conditions of heat or pressure, or both, used to accomplish bonding and the attainment of approximately maximum bond strength. Sometimes called aging time.

Coverage. The area over which a given quantity of material can be applied at a specified thickness.

Crazing. Fine cracks that may extend in a network on or under the surface of or through a layer of adhesive.

Delamination. The separation of layers in a laminate because of failure of the adhesive, either in the adhesive itself or at the interface between the adhesive and the adherend, or because of cohesive failure of the adherend.

Doctor blade. A scraper mechanism that regulates the amount of adhesive on the spreader roll or on the surface being coated.

Doctor roll. A roller mechanism that revolves at a different surface speed, or in an opposite direction, resulting in a wiping action for regulating the amount of adhesive supplied to the spreader roll.

Extender. A substance, generally having some adhesive action, added to an adhesive to reduce the amount of the primary binder required per unit area.

Filler. A relatively nonadhesive substance added to an adhesive to improve its working properties, permanence, strength, or other qualities.

Fillet. A rounded bead or a concave junction of sealing compound over or at the edges of structural members.

Glue. Originally, a hard gelatin obtained from hides, tendons, cartilage, bones, etc, of animals. Also, an adhesive prepared from this substance by heating with water. Through general use the term is now synonymous with the term *adhesive.*

Hardener. A substance or mixture of substances added to an adhesive to promote or control the curing reaction by taking part in it. The term is also used to designate a substance added to control the degree of hardness of the cured film.

Heat seal. To bond or weld a material to itself or to another material by the use of heat. This may be done with or without the use of adhesive, depending on the nature of the materials.

Joint. The location at which two adherends are held together.

Joint, lap. A joint made by placing one adherend partly over another and bonding together the overlapped portions.

Joint, scarf. A joint made by cutting away similar angular segments of two adherends and bonding the adherends with the cut areas fitted together.

Joint, starved. A joint that has an insufficient amount of adhesive to produce a satisfactory bond.

Laminate. A product made by bonding together two or more layers of material or materials.

Lamination. The process of preparing a laminate. Also, any layer in a laminate.

Mucilage. An adhesive prepared from a gum and water. Also in a more general sense, a liquid adhesive that has a low order of bonding strength.

Paste. A soft, plastic adhesive, such as that obtained by heating starch and water together.

Peel. To separate a bond of two flexible materials or of a flexible and a rigid material by pulling the flexible material from the joined surfaces.

Pick-up roll. A spreading device in which the roll for picking up adhesive revolves in a reservoir of adhesive.

Primer.   A coating applied to a surface, prior to the application of an adhesive, to improve the performance of the bond.

Set.   To convert an adhesive into a fixed or hardened state by chemical or physical action, such as polymerization, oxidation, vulcanization, gelation, dehydration, or evaporation of volatile constituents.

Sizing.   The process of applying a material on a surface in order to fill pores and thus reduce the absorption of the subsequently applied adhesive or coating or to otherwise modify the surface properties of the substrate to improve the adhesion. Also, the material used for this purpose.   The latter is sometimes called size.

Solvent cement.   To bond plastic materials the surfaces of which have been rendered tacky by the application of a solvent.

Spread.   The quantity of adhesive per unit joint area applied to an adherend.   It is often expressed in pounds of adhesive per thousand square feet of joint area. Single spread refers to application of adhesive to only one adherend of a joint. Double spread refers to application of adhesive to both adherends of a joint.

Stabilizer.   An ingredient used in the formulation of some adhesives, especially elastomers, to assist in maintaining the physical and chemical properties of the compounded materials at their initial values throughout the processing and service life of the material.

Storage life.   The period of time during which a packaged adhesive can be stored under specified temperature conditions and remain suitable for use.   Sometimes called shelf life.

Substrate.   A material upon the surface of which an adhesive-containing substance is spread for any purpose, such as bonding or coating.   A broader term than adherend.

Tack.   Stickiness of an adhesive.

Tack, dry.   The property of certain adhesives, particularly nonvulcanizing rubber adhesives, to adhere on contact to themselves at a stage in the evaporation of volatile constituents, even though they seem dry to the touch.   Sometimes called aggressive tack or contact bonding.

Warp.   A significant variation from the original, true, or plane surface.

Welding.   To unite by heating to a plastic or fluid state the surfaces of the parts to be joined, and then allowing them to flow together.

Wood failure.   The rupturing of wood fibers in strength tests on bonded specimens, usually expressed as the percentage of the total area involved that shows such failure.

Working life.   The period of time during which an adhesive, after mixing with catalyst, solvent, or other compounding ingredients, remains suitable for use.

General References

Adhesion and Adhesives, Fundamentals and Practice, symposia papers, Society of Chemical Industry, London; and John Wiley & Sons, Inc., New York, 1954.
"Adhesives Technology and Markets," *Adhesives Age* **2** (1), 34–47 (1959).
J. L. Been, "Room Temperature Bonding," *Mod. Plastics* **33** (7), 126 (1956).
J. J. Bikerman, *The Science of Adhesive Joints,* Academic Press, Inc., New York, 1961.

J. M. Black and R. F. Blomquist, "Relation of Polymer Structure to Thermal Deterioration of Adhesive Bonds in Metal Joints," *NASA Tech. Note D-108*, 1–35 (1959).

M. J. Bodnar, ed., *Symposium on Adhesives for Structural Applications*, Interscience Publishers, a division of John Wiley & Sons, Inc., New York, 1962.

R. S. Brookman, "Combining Vinyl and Cloth," *Adhesives Age* **2** (11), 30 (1959).

"Cementing, Welding, and Assembly of Plastics" and "Plastics as Adhesives," *Plastics Engineering Handbook of the Society of the Plastics Industry, Inc.*, 3rd ed., Reinhold Publishing Corp., New York, 1960, pp. 479–535.

N. A. deBruyne and R. Houwink, *Adhesion and Adhesives*, Elsevier Publishing Co., Amsterdam, 1951; 2nd ed., G. Salomon and R. Houwink, eds., in press.

H. W. Eickner, *General Survey of Data on the Reliability of Metal-Bonding Adhesive Processes*, Report No. 1862, U.S. Forest Products Laboratory, 1957.

D. D. Eley, ed., *Adhesion*, Oxford University Press, London, 1961.

G. Epstein, *Adhesive Bonding of Metals*, Reinhold Publishing Corp., New York, 1954.

W. H. Guttmann, *Concise Guide to Structural Adhesives*, Reinhold Publishing Corp., New York, 1961.

G. Koehn, "Design Manual on Adhesives," *Machine Design* **26**, 144–174 (April 1954).

H. H. Levine, "High-Temperature Structural Adhesives," *Ind. Eng. Chem.* **54** (3), 22–25 (1962).

J. C. Merriam, "Adhesive Bonding," *Mater. Design Eng. Manual* **162**, 113–128 (Sept. 1959).

H. R. Merriman and H. L. Goplen, "Research on Structural Adhesive Properties Over a Wide Temperature Range," *WADC Technical Report* **56-320**, Part I (1957), Part II (1958).

R. R. Meyers, "Roll Application of Adhesives," *Tappi* **46**, 745–750 (1963).

W. H. Neuss, ed., *Testing of Adhesives*, No. 26 in Tappi Monograph Series, Technical Association of the Pulp and Paper Industry, 1963.

H. A. Perry and R. H. Wagner, *Adhesive Bonding of Reinforced Plastics*, McGraw-Hill Book Co., Inc., New York, 1959.

"Problems in Wood Chemistry," *Proceedings of Session of the United Nations Food and Agricultural Organization, Jerusalem, April 8–13, 1956*, Weizmann Science Press of Israel, Jerusalem, 1957. Includes a discussion of adhesion and of wood adhesives with an extensive survey of the literature.

M. W. Riley, "Adhesive Bonding: State of the Art," *Mater. Design Eng.* **54** (7), 89–93 (1961).

H. Schonhorn, "Generalized Approach to Adhesion Via the Interfacial Deposition of Amphipathic Molecules," *J. Polymer Sci.* **1A**, 2343–2359, 3523–3536 (1963); *J. Appl. Polymer Sci.* **8**, 355–361 (1964).

L. H. Sharpe and H. Schonhorn, "Surface Energetics, Adhesion, and Adhesive Joints," No. 43 in *Advances in Chemistry* Series, American Chemical Society, Washington, D.C., 1964.

I. Skeist, *Handbook of Adhesives*, Reinhold Publishing Corp., New York, 1962.

E. Thelen, "Adherent Surface Preparation," *J. Appl. Polymer Sci.* **6**, 150–154 (1962).

S. S. Voyutskii, *Autohesion and Adhesion of High Polymers*, Vol. 4 of H. Mark and E. H. Immergut, eds., *Polymer Reviews* Series, Interscience Publishers, a division of John Wiley & Sons, Inc., New York, 1963.

P. Weiss, ed., *Adhesion and Cohesion*, Elsevier Publishing Co., Amsterdam, 1962.

H. P. Zade, *Heatsealing and High-Frequency Welding of Plastics*, Interscience Publishers, Inc., New York, 1959.

W. A. Zisman, "Influence of Constitution on Adhesion," *Ind. Eng. Chem.* **55**, 18–38 (1963).

EFFECT OF CHEMICAL CONSTITUTION

In ancient times it was found that solids could be made to adhere strongly by wetting each surface to be joined with a thin layer of a liquid that hardened or solidified gradually during contact. Until this century the selection of suitable adhesives and application techniques was an art depending on the use of glue formulations made from fish and animal products or cements made from inorganic slurries or solutions. The advent of synthetic polymeric adhesives having better and more reproducible properties promoted a wider use of adhesives in the past thirty years and increased interest in converting the art into a science. Despite the more advanced technology of today, there are still many unanswered questions about the principles underlying the mechanism of adhesion and there still is excessive empiricism in current application techniques. Guidance is also needed in the search for adhesives suitable under more extreme conditions.

In the past several decades important contributions to the subject of adhesion have included the nature of the molecular forces responsible for adhesion, the relation between adhesion and friction, the effects of not matching the physical properties of adhesive and adherend, the effect of voids and occlusions on the development of stress concentrations, and the role played by adsorbed films and inadequate wetting on joint strength as well as on the location of the fracture. Obviously, such diversity in the physical and chemical factors important in adhesion makes necessary an interdisciplinary approach in attempting to fully understand or to extend the subject. However, progress in understanding adhesion and adhesives is well evidenced in the excellent collections of papers edited by deBruyne and Houwink (1), Eley (2), and Weiss (3). This article emphasizes the physical chemical aspects of the subject, and especially the effect of chemical constitution on adhesion. See also Theory of Adhesive Joints, pp. 477–482.

Adhesion Between Dry Solids

Early in the nineteenth century it was recognized that adhesion is caused by a strong but extremely localized field of attractive force emanating from the surface of each solid and liquid. Much later it was learned that this field of force varies approximately as the inverse seventh power of the distance from each atom in the solid surface. In the absence of ions or strong permanent dipoles in or near the surface, the field intensity is generally negligible when the two surfaces of the attracting substances are separated by distances of the order of from 3 to 10 Å; as a result, only when there is a significant area of intimate contact between two solids can they be made to adhere perceptibly. When two dry, hard solids are pressed together little effort is required to pull them apart, and this occurs because (as Holm (4) and Bowden (5) showed) the *real area* of contact is generally such a small fraction of the *apparent area*. Reports that strong adhesion can occur between carefully polished contacting solids have been shown to result from a surface-tension effect caused by the presence of a thin layer of liquid between them.

7

Effect of Monomolecular Layers.   A film of oxide or of organic contamination only one molecule thick can *decrease* greatly the adhesion of solids.   If the oxide is removed by appropriate means, the solids can readily be made to adhere on contact. This fact is well exemplified by Anderson's simple experiment (6), in which two gold spheres are pressed together in a high vacuum.   Even under these circumstances they will not adhere.   However, if one gold ball is twisted about its axis while being pressed against the other, some of the adsorbed film and metal oxide is sheared off and areas of clean unoxidized gold are exposed; they come into intimate contact, and a strong joint is formed.   A striking demonstration of the effect of an organic monolayer on adhesion is McFarlane and Tabor's experiment (7) with a clean steel ball pressed against a piece of indium coated with a close-packed monomolecular film of lauric acid.   Although the monolayer prevented adhesion when close-packed, it lost that ability when the two solids were pressed together sufficiently to stretch the interface and increase by 5% the average area occupied by the adsorbed molecules of lauric acid.   Apparently, more asperities of the solids were able to make contact through the stretched film and so good adhesion developed.

Effect of Elastic Strains.   Bowden and Tabor (5) have proved that in addition to the influence of adsorbed films in decreasing the adhesion of dry solids, there is a loss of adhesion following elastic recovery, or release of elastic strains, on removing the load pressing two solids together.   In Figure 1 the location of areas of elastic deformation behind contacting asperities is indicated by the arrows.   Upon removing the load W, the elastic strains, which had developed at the base of each asperity, are released and the resulting internal stresses help to break the adhering junctions.   If these elastic stresses can be removed by annealing the two solids while load W is maintained, strong adhesive junctions will persist even after removing the load.   This mechanism explains the hot forging of metals at temperatures much below that required to cause any significant diffusion of metal atoms across the junctions (8).

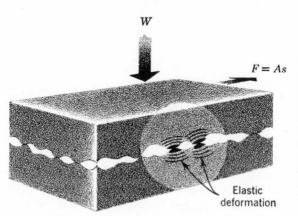

Fig. 1. Adhesion by solid–solid contact.   Adhesion or cold welding of two solids occurs every time there is intimate contact, at the points where plastic deformation occurs.   Friction is the force needed to break these bonds.   A load W helps create adhesion and also causes elastic deformation in the areas shown by arrows.   On removal of the load, the release of elastic strains causes internal stresses that act to break the bond. Annealing before removing the load would stabilize the bond.

Effect of Friction.   A wealth of convincing evidence that dry solids always adhere on contact has come from research on the nature of dry friction (or boundary friction).   Early investigations by Holm (4) and the now classic series of studies of Bowden, Tabor, and co-workers (5) have demonstrated that boundary friction is caused

Table 1. Effect of Constitution on Friction and Wettability of Halogenated Polyethylenes

Polymer	Structural formula	Static coefficient of friction	Critical surface tension, dyn/cm
poly(vinylidene chloride)	$\begin{bmatrix} & H & Cl \\ \sim\!\!\sim & C & C & \sim\!\!\sim \\ & H & Cl \end{bmatrix}_n$	0.90	40
poly(vinyl chloride)	$\begin{bmatrix} & H & Cl \\ \sim\!\!\sim & C & C & \sim\!\!\sim \\ & H & H \end{bmatrix}_n$	0.50	39
polyethylene	$\begin{bmatrix} & H & H \\ \sim\!\!\sim & C & C & \sim\!\!\sim \\ & H & H \end{bmatrix}_n$	0.33	31
poly(vinyl fluoride)	$\begin{bmatrix} & H & F \\ \sim\!\!\sim & C & C & \sim\!\!\sim \\ & H & H \end{bmatrix}_n$	0.30	25
poly(vinylidene fluoride)	$\begin{bmatrix} & H & F \\ \sim\!\!\sim & C & C & \sim\!\!\sim \\ & H & F \end{bmatrix}_n$	0.30	25
polytrifluoroethylene	$\begin{bmatrix} & F & F \\ \sim\!\!\sim & C & C & \sim\!\!\sim \\ & H & F \end{bmatrix}_n$	0.30	22
polytetrafluoroethylene	$\begin{bmatrix} & F & F \\ \sim\!\!\sim & C & C & \sim\!\!\sim \\ & F & F \end{bmatrix}_n$	0.04	18

by the adhesion of the two rubbing solids at the points of actual contact. In Figure 1 the region of the plastically deformed contacting asperities is indicated by the black areas. Such adhesion or *cold welding* occurs on contact of any two solids, making it necessary to apply a force F to overcome friction. To a first approximation, F is equal to the true area A of all the contacting asperities multiplied by the average shear strength s of the junctions. Adhesion between contacting solids is more readily made evident by sliding one over the other than by pulling them apart, because in measuring the frictional force there is no necessity of removing the internal stresses caused by the load in order to detect the existence of the effect.

An extensive literature now exists on the effect of varying the nature of the rubbing solids on their dry friction as well as on the influence of oxides or adsorbed films in lowering the friction (see FRICTION). Adsorbed monomolecular films of polar–nonpolar (or *amphipathic*) compounds can greatly decrease the friction between solids, and the physical adsorption of such compounds is the basic mechanism responsible for the operation of the oiliness additives used in lubrication technology. Such oil additives are most effective when they are adsorbed on the rubbing solids as solid monolayers.

In sliding together clean dry plastics, high friction (and even cold welding) is not uncommon. The range of possibilities and the effect of varying the chemical composition of the polymer are well illustrated on comparing in Table 1 the coefficient of static friction of two pieces of polyethylene with that of each structurally similar polymer formed by replacing some of the hydrogen atoms in the polymer with either fluorine or chlorine atoms (9). Whereas the coefficient of dry friction of polyethylene is about 0.3, it is lowered by fluorination, reaching 0.04 when fully fluorinated, as in the case of polytetrafluoroethylene; it is increased by progressive chlorination, rising to 0.9 with 50% chlorination, as in the case of poly(vinylidene chloride). As a matter of fact, the existence of such high friction and adhesion between some solid polymers has been utilized in processes for the frictional welding of plastics (see Bonding, below). Hence, in order to decrease friction and wear in processing, weaving, and application of the woven article, nearly all textile fibers must be treated or "finished" with lubricants.

Use of Liquids to Promote Adhesion

For over 40 years it has been known that when two flat, smooth, solid surfaces (the adherends) are separated by a thin layer of a liquid having a zero contact angle, strong adhesion will result. This effect arises from the liquid surface tension (γ_{LV°) and the fact that there will be a concave meniscus at the liquid–air interface (Fig. 2). If the area of contact of the liquid with the solid is circular and of radius R and thickness d, and if the liquid layer is thin enough so that the meniscus can be treated as a circular toroid of radius $r = d/2$, the Laplace equation of capillarity (9a) leads to equation 1. If $r << R$, then $p_L - p_A$ will be a large negative quantity and

$$p_L - p_A = \gamma_{LV^\circ} \left(\frac{1}{R} - \frac{1}{r} \right) \tag{1}$$

there will be a greater pressure p_A in the air outside the liquid than the value p_L inside; hence, the two plates will be pressed together under the pressure difference $p_A - p_L$. In short, a thin layer of a liquid that completely wets two flats solids can

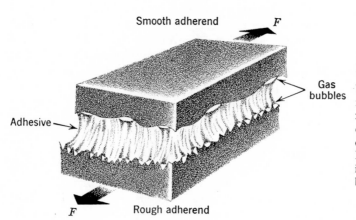

Smooth adherend   F

Gas bubbles

Adhesive

F   Rough adherend

Fig. 2. A realistic model of liquid–solid contact. Gas bubbles or voids prevent the bond from approaching theoretical strength. If roughness is fairly regular, bubbles can form along a plane and create a line of weakness. Deliberate roughening results in a random pattern of bubbles, with no single weak line.

serve as an adhesive. Equation 1 has been confirmed experimentally by Budgett (10), by Hardy (11), by Bastow and Bowden (12), and more recently by deBruyne (13,14). For example, Budgett found that two highly polished steel plates 4.5 cm in diameter, when completely wet by a film of paraffin oil having a surface tension of 28–30 dyn/cm, required a total force of about 20 kg to pull them apart. The calculated force is about 30 kg. Such a method of forming an adhesive joint has several obvious limitations: The resistance of the joint to shear stresses is determined solely by the viscosity of the liquid film, and hence only if the viscosity is very great could the shear strength of the joint be large; it would be necessary to prepare extremely well-fitted, smooth, solid surfaces to form a strong joint; and freedom from dust would be critically important.

If the contact angle of a liquid adhesive with each adherend is not large, a prompt adhesive action is obtained upon pressing them together, even though imperfectly fitted, until they are separated by only a thin liquid layer. Thereafter, a stronger and more useful joint will be formed if the viscosity of the liquid layer is increased greatly through any of various mechanisms, such as solvent permeation or evaporation (as in sticking on a postage stamp), polymerization, and cooling until solidification occurs. If properly designed, the resulting joint can have high resistance to both tensile and shearing stresses. However, although the process of forming an adhesive joint can be readily described, an analysis of the mechanisms involved and of the factors limiting the strength and utility of the joint is complex.

In their pioneering studies of adhesion in 1925, McBain and Hopkins (15) reported that "Any fluid which wets a particular surface and which is then converted into a tenacious mass by cooling, evaporation, oxidation, etc, must be regarded as an adhesive for that surface." McBain and Lee (16) soon afterward added the requirement that the adhesive must be able to deform during its solidification to release elastic stresses developed in the forming of the joint. Their work made evident the following three requirements for an adhesive: wetting, solidification, and sufficient deformability to reduce the buildup of elastic stresses in the formation of the joint. Hull and Burger (17) found the same rules necessary for good glass-to-metal joints. The second and third requirements have been fully investigated and reported. The first requirement has been studied much later by deBruyne (14) and Zisman (18–20).

Thermodynamic Conditions for the Wetting and Spreading of a Liquid on a Solid

Until recently, the essential physical and chemical variables controlling wetting and spreading have not been recognized nor controlled well enough to permit gathering reproducible and meaningful data. Despite the obvious importance of the subject, the effect of constitution of the liquid and solid on wetting and spreading has only begun to be explored.

Contact Angles. Basic to the subject of wettability is Young's concept (21) of the *contact angle* θ between a drop of liquid and a plane solid surface (Fig. 3). When $\theta > 0°$, the liquid is nonspreading. When $\theta = 0°$, the liquid is said to wet the solid completely, and it will then spread freely over the surface at a rate depending on the viscosity and surface roughness. Every liquid wets every solid to some extent; that is, $\theta \neq 180°$. In short, there is always some adhesion of any liquid to any solid. On a solid having uniform surface, the angle θ is independent of the volume of the liquid drop. Since the tendency for the liquid to spread increases as θ decreases, the con-

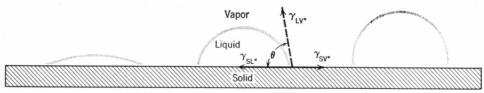

Fig. 3. Adhesion depends on wetting of the surfaces.   When $\theta = 0$, the liquid wets the surface completely and spreads freely.   Every liquid wets every surface to some extent—there is always some degree of adhesion.   Contact angle, θ, is a good inverse measure of wetting and spreadability, and hence of adhesion.   At the phase boundary, the three surface tensions shown must be in static equilibrium.

tact angle is a useful inverse measure of spreadability or wettability; obviously, cosine θ is a direct measure.   Methods of measuring θ have been reviewed (22).   They include profile-view measurements of the contact angle, using a goniometer eyepiece fitted to either a low-power telescope or microscope, and also the measurement of the height of a drop of known volume having the shape of a segment of a sphere.

Probably the oldest experimental problem in measuring contact angles has been the occurrence of large differences between the contact angle θ_A observed when a liquid boundary advances for the first time over a dry, clean, smooth surface and the value θ_R observed when the liquid boundary recedes from the previously wetted surface.   The most common cause of the differences observed is the effect of pores and crevices in the surface of the solid in trapping liquid as it flows over the surface. When the liquid recedes, the surface uncovered may have wet areas; hence, the receding contact angle is always lower than the advancing angle.   However, when sufficient care is exercised in preparing and handling smooth clean surfaces and when sufficiently pure liquids are used, no significant differences are found between the slowly advancing and receding contact angles (18–20,23–25).

Wenzel (26) showed in 1936 that the roughness r of a solid surface is related to the apparent, or measured, contact angle θ' between the liquid and the surface of the solid and the true contact angle θ (eq. 2).   Here a macroscopic roughness factor r

$$r = \frac{\cos \theta'}{\cos \theta} \qquad (2)$$

is defined as the ratio of the *true area* to the *apparent area* (or envelope) of the solid. This simple relation can readily be shown to be a consequence of the definition of r and the first two laws of thermodynamics (27,28).   Equation 2 cannot be ignored in practice because surfaces having $r = 1.00$ are rare; the nearest to such a smooth surface is that of freshly fire-polished glass.   Carefully machined or ground surfaces have values of r greater than 1.5.   There are several important consequences of Wenzel's equation.   Since r is always greater than unity, when $\theta < 90°$, equation 2 indicates $\theta' < \theta$.   Since most organic liquids exhibit contact angles of less than 90° on clean polished metals, the effect of roughening the metals is to make the apparent contact angle θ' between the drop and the envelope to the metal surface less than the true contact angle θ.   In other words, each liquid will appear to spread more when the metal is roughened.   When $\theta > 90°$, equation 2 indicates $\theta' > \theta$.   Since pure water makes a contact angle of from 105–110° with a smooth paraffin surface, the effect of roughening the surface tends to make θ' greater than 110°; values of 140° have been observed (27).

Reversible Work of Adhesion

Young (21) originated the idea 158 years ago that the three surface tensions γ_{SV°, γ_{SL}, and γ_{LV° existing at the phase boundaries of a drop of liquid at rest on a solid surface (see Fig. 3) must form a system in static equilibrium. The resulting relation was also derived by Sumner (29), and more rigorously later by Johnson (30), using an essentially thermodynamic argument; it led to substituting the various specific surface free energies in the system for the surface tensions. The resulting relation, the Young equation, is:

$$\gamma_{SV^\circ} - \gamma_{SL} = \gamma_{LV^\circ} \cos \theta \tag{3}$$

Here the subscripts SV° and LV° refer to the solid and liquid in equilibrium with the saturated vapor, respectively. The contact angle used here must be measured at thermodynamic equilibrium.

Almost a century ago Dupré (31) showed that W_A, the reversible work of adhesion per unit area of one liquid with another liquid (or with a solid), is related to the various specific surface free energies as follows:

$$W_A = \gamma_{S^\circ} + \gamma_{LV^\circ} - \gamma_{SL} \tag{4}$$

Here S° refers to the solid in a vacuum. If one is interested in the work W_{A*} required to pull the liquid away from the surface leaving the equilibrium-adsorbed film, it is given by the relation:

$$W_{A*} = \gamma_{SV^\circ} + \gamma_{LV^\circ} - \gamma_{SL} \tag{5}$$

Eliminating γ_{SL} from equations 3 and 5,

$$W_{A*} = \gamma_{LV^\circ} (1 + \cos \theta) \tag{6}$$

From an application of equation 4 to a liquid–liquid interface made up of two layers of the same liquid, the specific reversible work of cohesion W_c of the liquid is found to be simply

$$W_c = 2\gamma_{LV^\circ} \tag{7}$$

Surface Energies. Bangham and Razouk (32,33) in 1937 were the first to recognize that the surface energy change resulting from the adsorption of the vapor of the liquid on the surface of the solid cannot be generally neglected; they derived the following equation for W_A from equations 4, 5, and 6:

$$W_A = (\gamma_{S^\circ} - \gamma_{SV^\circ}) + \gamma_{LV^\circ} (1 + \cos \theta) \tag{8}$$

Here γ_{SV° is the specific surface free energy of the solid immersed in the saturated vapor of the liquid; hence $\gamma_{S^\circ} - \gamma_{SV^\circ}$ is the specific surface free energy decrease on immersion of the solid in the saturated vapor of the liquid. The symbol f_{SV° will be used to represent this free energy change, ie

$$f_{SV^\circ} = \gamma_{S^\circ} - \gamma_{SV^\circ} \tag{9}$$

Thus equation 8 can be written:

$$W_A = f_{SV^\circ} + \gamma_{LV^\circ} (1 + \cos \theta) \tag{10}$$

Hence, from equation 6

$$W_A - W_{A*} = f_{SV^\circ} \tag{11}$$

Bangham and Razouk also pointed out that if the surface concentration of the adsorbed vapor from the liquid is Γ and the chemical potential is μ then equation 12 holds,

$$\Gamma = -\frac{\partial \gamma}{\partial \mu} \tag{12}$$

or

$$\gamma_{S^\circ} - \gamma_{SV^\circ} = \int_{p_i = 0}^{p = p_0} \Gamma \, d\mu \tag{13}$$

where p is the vapor pressure and p_0 the saturated vapor pressure. In the special case where the vapor behaves like a perfect gas, $d\mu = RT \, d \, (\ln p)$, so that

$$f_{SV^\circ} = \gamma_{S^\circ} - \gamma_{SV^\circ} = RT \int_p^{p_0} \left(\frac{\Gamma}{p}\right) dp \tag{14}$$

Hence, f_{SV° will always have a positive value, and in general $W_A > \gamma_{LV^\circ} \, (1 + \cos \theta)$, or $W_A > W_{A*}$.

Cooper and Nuttall (34) appear to have originated the well-known conditions for the spreading of a liquid substance b on a solid (or liquid) substance a:

$$S > 0 \quad \text{for spreading} \qquad S < 0 \quad \text{for nonspreading} \tag{15}$$

where

$$S = \gamma_a - (\gamma_b + \gamma_{ab}) \tag{16}$$

or in the above notation for a solid–liquid system

$$S = \gamma_{S^\circ} - (\gamma_{LV^\circ} + \gamma_{SL}) \tag{17}$$

Harkins soon afterward developed and applied these relations more fully in a series of papers (35–38), identified S as the *initial spreading coefficient*, and derived the relations

$$S = W_A - W_c \tag{18}$$

$$S = -\frac{\partial G}{\partial \sigma} \tag{19}$$

where G is the free energy of the system and σ the surface area.

Assuming there is no surface electrification, equations 3–19 are the basic thermodynamic relations for wetting and spreading under equilibrium conditions. In recent years, Skinner, Savage, and Rutzler (39–41) and Deryagin (42) have reported very interesting experiments on the occurrence of electrification during the breaking of adhesive joints with metals and have argued that there is always an important electrostatic contribution to adhesion. Their conclusions relate to the rapid breaking of adhesive joints; they may not be applicable to failure under slowly applied stresses. It is suggested their results are indicative of a consequence of adhesion rather than a cause. Other limitations of the electrostatic theory have been well summarized recently by Voyutskii (43,44).

An informative approximation can be derived from equation 17 for the case of an organic liquid spreading upon an organic solid surface, since it is then reasonable to assume that γ_{SL} is negligibly small in comparison with γ_{LV°. Therefore

$$S = \gamma_{S^\circ} - \gamma_{LV^\circ} \tag{20}$$

$$\gamma_{S^\circ} > \gamma_{LV^\circ} \quad \text{for spreading} \tag{21}$$

Hence in all such systems, when spreading occurs, the specific surface free energy of the liquid is usually less than that of the solid.

Wettability can also be measured by the reversible work of adhesion W_A or by the heat of wetting per unit area h_{SL}. However, the small change in energy involved in contacting most solids and liquids necessitates using finely divided solids having large surface areas per gram. Preparation of such materials is difficult with many soft solids of interest. The grinding operation also introduces new complications. In addition to the known sensitivity of such measurements to traces of impurities, there must be added the effect of the presence of strains, sharp edges, holes, and other surface imperfections. Such problems have plagued all investigators.

It is convenient to name the two extremes of the specific surface free energies of solids. The surface free energies of all liquids (excluding the liquid metals) are less than 100 erg/cm² at ordinary temperatures. Hard solids have surface free energies ranging from around 5000 to about 500 erg/cm², the values being higher the greater the hardness and the higher the melting point. Examples of such solids are the ordinary metals, metal oxides and nitrides, silica, glass, ruby, and diamond. Soft organic solids have much lower melting points and their specific surface free energies are generally under 100 erg/cm². Examples are waxes, most solid organic polymers, and in fact most organic compounds. Therefore, solids having high specific surface free energies may be said to have high-energy surfaces, and those having low specific surface free energies have low-energy surfaces (23).

Because of the comparatively low surface energies of organic and most inorganic liquids, one would expect them to spread freely on solids of high surface energy since there would result a large decrease in the surface free energy of the system; and this is most often found to be true. But the surface free energies of such liquids are comparable to those of low-energy solids; hence, systems showing nonspreading should be and are most commonly found among such combinations.

Effect of Constitution on Wetting

Systematic studies by Zisman and co-workers of the equilibrium contact angles of a wide variety of pure liquids on low-energy solid surfaces (20,45,46) and on high-energy solid surfaces (47,48) have revealed many interesting regularities. Among the solids studied were smooth organic crystals and polymeric solids, high-energy surfaces (such as high-melting metals and glass), and a number of such solid surfaces that had been modified by the adsorption of a monolayer of oriented organic molecules. A general rectilinear relation was established empirically (18,20,24,25,42) between θ and the surface tension γ_{LV° for each homologous series of organic liquids. This led to the development of γ_c, the *critical surface tension* of wetting, for each homologous series, as defined by the intercept of the horizontal line $\cos\theta = 1$ with the extrapolated straight line plot of $\cos\theta$ vs γ_{LV°.

Low-Energy Surfaces. The regularities in the contact angles exhibited by pure homologous liquids on low-energy surfaces are illustrated by the data of Figure 4 for the family of liquid *n*-alkanes on each of several types of fluorinated solid surfaces. Curve A is a plot for smooth, clean polytetrafluoroethylene (Teflon); the lower the surface tension of the alkane liquid, the larger is $\cos\theta$, and the more wetted the sur-

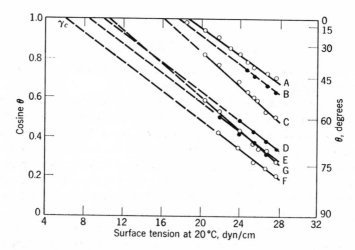

Fig. 4. Contact angles formed by a series of liquid *n*-alkanes on various fluorinated low-energy solid surfaces. The regularity of the data is striking. A. Polytetrafluoroethylene (Teflon). B. F.E.P. Teflon. C. Polyperfluoropropylene. D. Perfluorobutyric acid monolayer. E. Perfluorocaprylic acid monolayer. F. Perfluorolauric acid monolayer. G. Poly(1,1-dihydroperfluorooctyl methacrylate).

face. For all values of the surface tension of the liquids below a critical value γ_c, the contact angle is zero. Curve B presents recent data (49) for the new copolymer of tetrafluoroethylene and hexafluoropropylene (F.E.P. Teflon), and curve C is for a still newer material, polyperfluoropropylene (50). Curves D, E, and F describe the wetting behavior of three completely wettable, high-energy surfaces (clean, smooth platinum), each modified previously by the adsorption of an oriented, close-packed, monomolecular layer of a fully fluorinated (or perfluoro) fatty acid (51,52). The similarities in these graphs of the wetting properties are striking.

From the $\cos\theta = 1$ intercepts on Figure 4, it is evident that γ_c has a value of about 18.5 dyn/cm for the *n*-alkanes on the surface of Teflon. Values of about 17 and 15 dyn/cm are obtained from curves B and C; that is, the introduction of the perfluoromethyl group as a side chain in the polymer reduces γ_c, the reduction becoming greater the higher the surface concentration of exposed —CF_3 groups. An adsorbed, close-packed monolayer of a perfluoro fatty acid (curves D, E, and F) is an extreme example of such a surface (51,52). The values of γ_c for such surfaces are therefore much lower than for surfaces comprised only of —CF_2— groups. The closer the packing of the aliphatic chains of the adsorbed molecules, the closer the packing of the exposed terminal —CF_3 groups, and hence the lower γ_c. Thus the value for a condensed monolayer of perfluorolauric acid (curve F) is only 6 dyn/cm, and this surface is the least wetted by the alkanes (or by any other liquids) of any solid surface yet encountered.

Even when $\cos\theta$ is plotted against γ_{LV° for a variety of nonhomologous liquids, the graphical points lie close to a straight line or tend to collect around it in a narrow rectilinear band. On some low-energy surfaces this band exhibits curvature for values of γ_{LV° above 50 dyn/cm (23–25,53). But in those cases the curvature has been shown to result because weak hydrogen bonds form between the molecules of liquid and those in the solid surface. Thus curvature of the graph is most likely to happen with liquids of high surface tension, because these are always hydrogen-donating liquids. In general, the graph of $\cos\theta$ vs $\gamma_{LV\circ}$ for any low-energy surface is always a straight line (or a narrow rectilinear band) unless the molecules in the solid surface form hydrogen bonds or otherwise strongly associate with the liquid (53).

Critical Surface Tension. When rectilinear bands are obtained in this type of graph, the intercept of the lower limb of the band at $\cos\theta = 1$ is chosen as the critical surface tension γ_c of the solid. Although this intercept is less precisely defined than the critical surface tension of an homologous series of liquids, nevertheless it is an even more useful parameter *because it is a characteristic of the solid only*. It has been found to be an empirical parameter that varies with solid surface composition very much as would be expected of $\gamma_{S\circ}$, the specific free energy of the solid. The widespread occurrence of the rectilinear relationship between $\cos\theta$ and $\gamma_{LV\circ}$ in the now large body of experimental data, and the fact that these graphs rarely cross, had made it possible to use γ_c to characterize and compare the wettabilities of a variety of low-energy surfaces. By comparing the wetting properties of structurally homologous or analogous solids, such as unbranched polyethylene versus its chlorinated or fluorinated analogs (see last column of Table 1), it is possible to learn how the surface constitution of such solids affects the values of γ_c.

In Table 2 will be found the values of γ_c obtained from such studies (18–20,46) of the contact angles of a number of well-defined, low-energy, solid surfaces. In the first column is given the constitution of the atoms or organic radicals in the solid surface arranged in the order of increasing values of γ_c. Salient features of Table 2 deserve a brief discussion. The surface of lowest energy ever found (and hence having the lowest γ_c) is that comprised of closest packed —CF$_3$ groups (51,52). Replacement of a single fluorine atom by a hydrogen atom in a terminal —CF$_3$ group

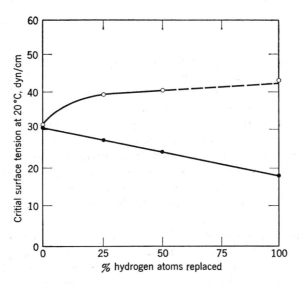

Fig. 5. Progressive halogen substitution of polyethylene-type surfaces. The critical surface tension of wetting is sensitive to both degree and type of substitution. ○ By chlorine. ● By fluorine.

more than doubles γ_c (53). A parallel and regular increase in γ_c has been observed with progressive replacement of fluorine by hydrogen atoms in the surfaces of bulk polymers. The data for polytetrafluoroethylene, polytrifluoroethylene, poly(vinylidene fluoride), and poly(vinyl fluoride) are listed in the order of increasing values of γ_c; however, this is also the order of decreasing fluorine content. A plot of γ_c against the atom percent replacement of hydrogen in the monomer by fluorine results in a straight line (Fig. 5).

Studies of the wetting of various types of hydrocarbon surfaces have revealed that the lowest values of γ_c are to be found in a surface comprising close-packed, oriented, methyl groups (Table 2). The low value of γ_c of 22 dyn/cm results when the

Table 2. Critical Surface Tensions of Low-Energy Surfaces[a]

Surface constitution	γ_c, dyn/cm
Fluorocarbon surfaces	
—CF$_3$	6
—CF$_2$H	15
—CF$_3$ and —CF$_2$—	17
—CF$_2$—CF$_2$—	18
—CF$_2$—CFH—	22
—CF$_2$—CH$_2$—	25
—CFH—CH$_2$—	28
Hydrocarbon surfaces	
—CH$_3$ (crystal)	20–22
—CH$_3$ (monolayer)	22–24
—CH$_2$—CH$_2$—	31
∴CH⋯ (benzene ring edge)	35
Chlorocarbon surfaces	
—CClH—CH$_2$—	39
—CCl$_2$—CH$_2$—	40
=CCl$_2$	43

[a] At 20°C.

methyl groups are in the close-packed array found in the easiest cleavage plane of a single crystal of a higher paraffin, such as n-hexatriacontane (25). The less closely packed arrangement found in a condensed adsorbed monolayer of a high-molecular-weight fatty acid is characterized by a value of γ_c of between 22 and 24 dyn/cm (18). The great sensitivity of the contact angle (and hence of γ_c) to such subtle changes in the packing of the methyl groups comprising the surface of the solid is remarkable, and it has much significance in technological aspects of wetting and adhesion. It should be noted that the transition from a surface comprised of —CH$_3$ groups to one of —CH$_2$— groups results in an increase in γ_c of some 10 dyn/cm; this is to be compared with the increase of 12 dyn/cm, observed in going from a surface of —CF$_3$ to one of —CF$_2$— groups. The presence of aromatic carbon atoms in the hydrocarbon surface also serves to increase γ_c. Thus the introduction of a significant proportion of phenyl groups in the surface in going from polyethylene to polystyrene raises γ_c from 31 to 33 dyn/cm. A further increase to 35 dyn/cm results when the surface is composed solely of phenyl groups edge on, as in the cleavage surface of naphthalene or anthracene single crystals (45).

In Figure 5 values of γ_c for polyethylene, poly(vinyl chloride), and poly(vinylidene chloride) (54) are plotted against the % atom replacement of hydrogen by chlorine.

Although the introduction of the first chlorine atom in the monomer causes γ_c to rise from 31 to 39 dyn/cm, the addition of a second chlorine only increases γ_c to 40 dyn/cm. There are striking differences, therefore, in the effects on γ_c observed with fluorine and chlorine replacement of hydrogen, both as to the direction of the change and the effect of progressive halogenation. Although polytetrachloroethylene has not been prepared, an organic coating with an outermost surface comprised of close-packed covalent chlorine atoms may be made by adsorbing a condensed, oriented monolayer of perchloropenta-2,4-dienoic acid ($CCl_2=CCl-CCl=CCl-COOH$) on a clean polished surface of glass or platinum (54). Not only is the graph of $\cos\theta$ vs γ_{LV° for such a surface quite similar to those of the above-mentioned chlorinated polyethylenes, but the corresponding value of γ_c (43 dyn/cm) is shifted in the appropriate direction (that is, to higher values of γ_c). Extrapolation of the line defined by the experimental

Table 3. Critical Surface Tensions of Polymeric Solids[a]

Polymeric solid	γ_c, dyn/cm
poly(1,1-dihydroperfluorooctyl methacrylate)	10.6
polyperfluoropropylene	16.2
polytetrafluoroethylene	18.5
polytrifluoroethylene	22
poly(vinylidene fluoride)	25
poly(vinyl fluoride)	28
polyethylene	31
polychlorotrifluoroethylene	31
polystyrene	33
poly(vinyl alcohol)	37
poly(methyl methacrylate)	39
poly(vinyl chloride)	40
poly(vinylidene chloride)	40
poly(ethylene terephthalate)	43
poly(hexamethylene adipamide)	46

[a] At 20°C.

points for the two chlorinated polymers in Figure 5 to the value of γ_c for 100% hydrogen replacement indicates a value of 42 dyn/cm. Thus the hypothetical polytetrachloroethylene surface should have a critical surface tension of wetting of 42 dyn/cm, which is only 1 dyn/cm less than the experimental value found for the adsorbed perchloropentadienoic acid monolayer. This shows how closely the wetting properties of the latter surface approximate those of a fully chlorinated polymeric solid surface.

Results of wettability studies on clean, smooth, plasticizer-free polymeric solids of general interest have been summarized in Table 3. Nylon-6,6, with its many exposed amide groups, has the highest values of γ_c of the common plastics reported (55). Since γ_c for all the polymers of Table 3 is well below the surface tension of water (72.8 dyn/cm), all are hydrophobic.

High-Energy Surfaces. In order to understand the wettability properties of high-energy surfaces, two new surface-chemical problems had to be solved. The first was encountered when it was found that liquids such as 1-octanol, 2-octanol, 2-ethylhexanol, trichlorodiphenyl (mixture of isomers), and tri-*o*-cresyl phosphate exhibited appreciable contact angles on clean, hydrophilic, high-energy surfaces, such as platinum, stainless steel, glass, fused silica, and α-Al$_2$O$_3$, no matter what extremes of purification were used. A lengthy investigation (47,48) revealed that each liquid was nonspreading because the molecules of the liquid were adsorbed on the solid to form a film whose critical surface tension of wetting was less than the surface tension of the liquid itself. In short, each liquid was unable to spread upon its *own* adsorbed and oriented monolayer; hence, such liquids were named *autophobic* liquids. The second problem was to explain why all pure liquid esters spread completely upon the metals studied, but only some spread on glass, silica, and α-Al$_2$O$_3$. It was found (47) that the cause of these differences in spreadability was ester hydrolysis, which occurred immediately after the liquid ester adsorbed upon hydrated surfaces, such as those of glass, fused silica, and α-Al$_2$O$_3$. This result is not unreasonable, since the polar group of the ester would be expected to adsorb in immediate contact with the solid surface unless prevented by steric hindrance and since in the surface the molecules of the water of hydration (being oriented) should be more effective in causing hydrolysis than bulk water. Through surface hydrolysis two fragments of the ester result. The fragment that has a greater average lifetime of adsorption remains and eventually coats the surface with a close-packed monolayer of that molecular species. Eventually the surface becomes blocked or poisoned by the coating of the hydrolysis product, and the hydrolysis reaction ceases. Hence the volume concentration of hydrolyzed ester is so small that it cannot be observed by applying ordinary analytical methods. When the resulting monolayer has a critical surface tension of wetting less than the surface tension of the ester, nonspreading behavior is observed; that is, the ester is unable to spread upon the adsorbed film of its own hydrolysis product. Dozens of pure esters having a great variety of structures were studied, and in every instance of nonspreading on glass, fused silica, and α-Al$_2$O$_3$, a similar explanation was found.

Upon applying this background of knowledge it becomes possible to explain the nearly universal spreading properties of the polydimethylsiloxanes. These liquids spread on all high-energy surfaces because the surface tensions of 19–20 dyn/cm (18,47) are always less than the critical surface tensions of wetting of their own adsorbed films. This follows because an adsorbed close-packed monolayer of such a liquid has an outermost surface of methyl groups which are not as closely packed as the methyl groups in a single crystal of a paraffin. Since γ_c of hexatriacontane is about 22 dyn/cm (25), the value of γ_c for the silicone monolayer must exceed that value; actually it is about 24 dyn/cm (19). Hence γ_{LV° is below γ_c, and the polydimethylsiloxanes cannot be autophobic (47).

Similarly it is possible to explain the spreading properties of the liquid paraffins. The critical surface tension of polyethylene of 31 dyn/cm (25) must be about the same as that of a monolayer of paraffin molecules adsorbed lying flat on a metal (45). Since the surface tensions of liquid aliphatic hydrocarbons are always less than 30 dyn/cm and hence less than γ_c, such liquids are always able to spread on their own adsorbed films. Thus the liquid paraffins cannot be autophobic.

Many liquid compounds release polar decomposition products able to adsorb and form low-energy surfaces; in fact, the nonspreading property can be produced in nearly all pure liquids by the addition of a minor concentration of a suitable adsorbable polar compound (18–20). Such addition agents in oils make them adsorb oleophobic monolayers upon contact with suitable high-energy surfaces.

Generalizations about Wetting and Spreading on Solids

The results of these investigations can be generalized as follows (18–20): Every liquid having a low specific surface free energy always spreads freely on a specularly smooth, clean, high-energy surface at ordinary temperatures unless the film adsorbed by the solid converts it into a low-energy surface having a critical surface tension less than the surface tension of the liquid. Because of the highly localized nature of the forces between each solid surface and the molecules of a liquid and also between the molecules of each liquid, a monolayer of adsorbed molecules is always sufficient to give the high-energy surface the same wettability properties as the low-energy solid having the same surface constitution.

These conclusions are based upon experiments. They also are good evidence of the extreme localization of the attractive fields of force around solid surfaces, especially those comprised of covalent-bonded atoms. The influence of the intermolecular fields of force, such as the London dispersion forces (56), are known to become unimportant in very simple molecules at a distance of separation of only a few atom diameters (57). In dealing with solids and liquids comprised of such molecules, one would expect that there would be little contribution to the force of adhesion by atoms not in the surface layers. However, substances made from more complex molecules present theoretical problems solely because of the mathematical difficulties involved. Recent investigations have revealed that when either ions or large, uncompensated, permanent dipoles were located in the outermost portion of the surface of the solid or liquid, the residual field of force of the surface was much less localized, and hence wetting was influenced by the constitution of the solid below the atoms in the surface (58,59). Langmuir (60–64) called attention many years ago to the extreme localization of surface forces encountered in observing the mechanical properties of insoluble organic monolayers and often referred to this concept as "the principle of independent surface action." Studies of wetting by Shafrin and Zisman (20,58,59) have demonstrated that although there are understandable exceptions to the principle, it is very commonly true.

Reversible Work of Adhesion of Liquids to Solids

High-Energy Surfaces. Bangham and Razouk's equations (65) and (66) for the reversible work of adhesion have been used by Boyd and Livingston (67) and Harkins and co-workers (68,69) to calculate W_A from experimental values of f_{SV^o} for various liquids on a number of metallic and nonmetallic highly adsorptive, finely divided high-energy surfaces. In each case f_{SV^o} was computed approximately from the measured adsorption isotherm of the vapor on the solid. It was concluded that in every instance f_{SV^o} is of the same magnitude as W_{A*} and so cannot be neglected in the calculation of W_A.

Tabor (70) has pointed out that from equation 10 it follows that W_{A*} for any system having $\theta = 0$ will be equal to or greater than twice the surface tension of the liquid. A simple calculation shows that if the field of force emanating from the solid surface is assumed to vanish in about 3 Å and if the surface tension of the liquid adhesive is 30 dyn/cm, then the average tensile strength of the adhesive joint is 2000 kg/cm². This value is much greater than the tensile strength of common adhesives. For example, Kraus and Manson (70a) have obtained a tensile strength of 183 kg/

cm² for polyethylene. Therefore, the joint must break by cohesive failure (that is, failure in the bulk phase of the adhesive). Since the correction term $f_{SV\circ}$ in equation 10 simply makes the energy of adhesion even greater than $2\gamma_{LV\circ}$, it is apparent that when the adhesive makes a zero contact angle with the adherend, the theoretical adhesive strength will always be much more than the observed tensile or shear strength of the adhesive.

Low-Energy Surfaces. Unfortunately, reliable data are not available on the values of $f_{SV\circ}$ for well-defined, smooth, low-energy, solid surfaces. But one should not assume from the results with high-energy surfaces that the value of $f_{SV\circ}$ will also be an important correction term in equation 10 for low-energy solids. On the contrary, there is good experimental evidence that whenever a liquid exhibits a large contact angle on a solid, there is negligible adsorption of the vapor (19). Recent measurements by Bewig and Zisman (71) using a new contact potential method of studying adsorption have demonstrated that negligible adsorption of most vapors occurs on smooth, clean surfaces of polytetrafluoroethylene. Adsorption measurements with a McBain-Baker balance over the entire vapor pressure range of p/p_0 up to 1.0, by Martinet (72), and a more recent investigation by Graham (73) have led to the conclusion that the vapor adsorption at ordinary temperatures for the many substances studied is a small fraction of a monolayer. Hence $f_{SV\circ}$ must be a very small term compared with W_{A*} for all liquids having $\gamma_{LV\circ}$ greater than γ_c.

The same conclusions can be extended for the same temperatures to other low-energy solids, such as polyethylene, polystyrene, poly(vinyl chloride), etc, and to all liquids having surface tensions greater than the value of γ_c of each of these solids. If a substantial fraction of a condensed monolayer of any of these vapors were to adsorb on the solid surface, the surface constitution (and hence γ_c) of the solid surface would necessarily be transformed to that of the adsorbed film, and the wetting properties would have to become those of a higher energy surface. Therefore, it can be concluded that $f_{SV\circ}$ must be negligible for liquids having surface tensions considerably larger than γ_c; it presumably becomes a more significant correction for liquids having values of $\gamma_{LV\circ}$ close to or less than γ_c. Furthermore, when $\theta >> 0$, the liquid on a low-energy solid surface will be in equilibrium with a small fraction of an adsorbed monolayer (19,71). For any high-energy surface that has been converted to a low-energy surface by the adsorption of a suitable condensed organic monolayer, the same conclusions about the negligible value of $f_{SV\circ}$ apply as those given in the preceding paragraph, with the one reservation that $f_{SV\circ}$ may become more significant if the molecule of the liquid is small enough to penetrate readily through the condensed monolayer and so adsorb on the high-energy surface beneath.

Therefore, for low-energy solid surfaces it may be said that: (a) liquids having $\gamma_{LV\circ}$ much greater than γ_c will have W_A essentially equal to W_{A*}; (b) as $\gamma_{LV\circ}$ closely approaches γ_c but exceeds it, $W_A - W_{A*}$ may become more significant; and (c) liquids having $\gamma_{LV\circ}$ less than or equal to γ_c, may have appreciable values of $W_A - W_{A*}$. It can also be concluded that the difference $W_A - W_{A*}$ will not be negligible for high-energy surfaces unless both the vapor pressure and adsorptivity are very low or unless the liquid on contact with the solid lays down a low-energy adsorbed film; in such cases rules (a), (b), and (c) will apply.

Adhesion to a Planar Surface. In dealing with the adhesion of a liquid to a planar, nonporous, solid surface, according to equation 6, W_{A*} is given by the equation $W_{A*} = \gamma_{LV\circ} (1 + \cos \theta)$. However, experiments show that in any homologous series of liquid compounds and for all values of $\gamma_{LV^\circ} > \gamma_c$, the contact angle is related to γ_{SV° by the straight line equation:

$$\cos \theta = a - b\gamma_{LV^\circ} \tag{22}$$

Since γ_{LV° approaches γ_c as θ approaches zero, equation 22 can be written

$$\cos \theta = 1 + b(\gamma_c - \gamma_{LV^\circ}) \tag{23}$$

Upon elimination of $\cos \theta$ between equations 6 and 23, there results (19):

$$W_{A*} = (2 + b\gamma_c)\gamma_{LV^\circ} - b\gamma_{LV^\circ}^2 \tag{24}$$

This is the equation of a parabola with the concave side toward the surface tension axis; it has a maximum value of W_{A*} occurring at

$$\gamma_{LV^\circ} = \frac{1}{b} + \left(\frac{1}{2}\right)\gamma_c \tag{25}$$

Finally, the maximum value of W_{A*} is given by

$$W_{A*} = \left(\frac{1}{b}\right) + \gamma_c + \left(\frac{1}{4}\right)b\gamma_c^2 \tag{26}$$

For example, with a sufficiently smooth surface of polyethylene $\gamma_c = 31$ and $b = 0.026$ (25); the maximum of W_{A*} occurs at $\gamma_{LV^\circ} = 54$ dyn/cm and is about 76 erg/cm².

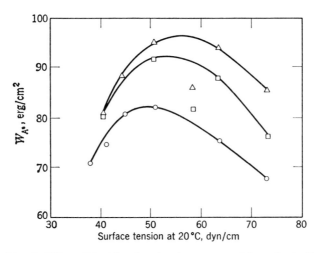

Fig. 6. Effect of liquid surface tension on reversible work of adhesion (W_{A*}) for polyethylene-type polymers with chlorine substitution. △ Poly(vinylidene chloride). □ Poly(vinyl chloride). ○ Polyethylene.

It is of interest to compute the value of W_{A*} for the many liquid–solid combinations reported in the past. Figures 6 to 9 show how W_{A*} varies as a function of γ_{LV°. In every case there results a parabolic curve with a well-defined maximum. However, in a few instances, such as in the case of polytrifluoroethylene (Fig. 7), the decrease of W_{A*} after reaching the maximum value is greatly moderated by the effect

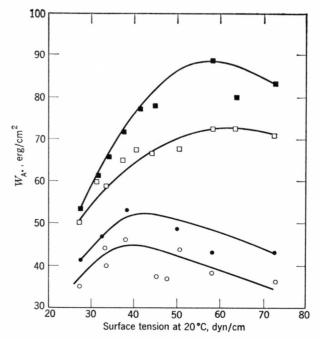

Fig. 7. Effect of liquid surface tension on reversible work of adhesion (W_A*) for various polymers with fluorine substitution. ■ Poly(vinylidene fluoride). □ Polytrifluoroethylene. ● Polyperfluoropropylene. ○ Poly-(1,1-dihydroperfluorooctyl methacrylate).

of the hydrogen-bonding action of the liquids of high surface tension, that is, water, glycerol, and formamide, each of which is an effective hydrogen-donating compound. Similar effects are seen in the curve of Figure 9 for the close-packed monolayers terminated by —CF_2H and —CF_3 groups. In the latter case, if the data points for the hydrogen-donating liquids are excluded, the curve is seen to form a parabola with its maxium occurring at about 40 dyn/cm. It is also seen that the maximum values of W_A* for —CH_3, —CF_2H, and —CF_3 coated surfaces are approximately 70, 65, and 44 erg/cm^2, respectively. Inspection of Figures 6 to 9 also reveals that the greatest variation in W_A* among all the low-energy solid surfaces reported is at the most threefold.

Adhesion to a Nonplanar Surface. Real solid surfaces are neither flat nor free of pores and crevices, and no treatment of adhesion is adequate if it disregards the realities of surface structure. Each adherend has a true surface area that is r times greater than the apparent or envelope area; hence the work of adhesion should be expected to be r times greater than that for the apparent surface area. However, the larger the contact angle the more difficult it becomes to make the liquid flow over the surface of each adherend to fill completely every crevice and pore in the surface. More often there are air pockets or voids in hollows and crevices. Such difficulties with the formation of gas bubbles and pores are, of course, greatly amplified in dealing with viscous adhesives that solidify shortly after being applied to the joint. Hence in practice the true value of W_A* lies somewhere between the value obtained from equation 6 and r times that value. Where there are accessible pores, crevices, and capillaries in

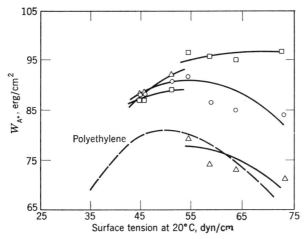

Fig. 8. Effect of liquid surface tension on reversible work of adhesion (W_A*) for several common plastics. □ Nylon. ○ Poly(ethylene terephthalate). △ Polystyrene.

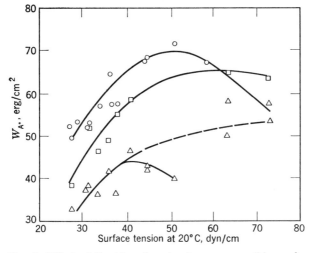

Fig. 9. Effect of liquid surface tension on reversible work of adhesion (W_A*) for surfaces with an adsorbed condensed layer of molecules terminating with the groups shown. □ CF_2H. ○ CH_3. △ CF_3.

the surface of the adherend, the liquid adhesive will penetrate to some extent and so increase adhesion; the process ceases when an adequate supply of liquid is no longer available or when the liquid viscosity increases too much or solidification occurs. Obviously, in order to obtain the maximum adhesion, the adhesive should penetrate and fill each capillary.

As an approximation it may be assumed that the capillary rise h given by equation 27 can be used (R is the equivalent radius of the capillary, $k = 2/981$, and ρ is the

$$h = \frac{k\gamma_{\text{LV}\circ} \cos \theta}{\rho R} \tag{27}$$

density of the liquid) and also that the liquid still wets the capillary wall according to the cos θ vs γ_{LV° relation of equation 22. Eliminating cos θ from the two equations, there results the following equation for the parabola:

$$h = \frac{k\gamma_c}{\rho R}(b + \gamma_c + 1) - \frac{(bk)}{R\rho}\gamma_{LV^\circ}^2 \qquad (28)$$

Evidently h has a maximum when

$$\gamma_{LV^\circ} = \frac{1}{2}\left(\gamma_c + \frac{1}{b}\right) \qquad (29)$$

For example, in the case of smooth polyethylene, the maximum capillary rise will occur when $\gamma_{LV^\circ} = 1/2$ (31 + 38.4) = 34.7 dyn/cm. Hence, the maximum rise occurs when γ_{LV° is 3.7 dyn/cm more than γ_c. It is questionable that the rise in the fine pores and crevices connected to the interface can be treated precisely in the above way in practice; it is also doubtful that the contact angle is uniform in the interior of the crevices and pores. However, this simple analysis reveals again that W_{A^*} may go through a maximum as γ_{LV° increases even when a porous interfacial surface is involved (19).

Effects of Solidification of the Liquid Adhesive

When a liquid adhesive solidifies, the reversible work of adhesion of the adhesive to the adherend would still be close to the value computed in the preceding calculations for the adhesive in the liquid state if it were not for the development of stress concentrations. This conclusion follows from the highly localized nature of the attractive field of force causing adhesion. Since this attractive force is effective little more than the depth of one molecule in both the adhesive and the adherend, it will be unaffected by changes of state so long as allowance is made for any resulting alteration in the surface density or molecular orientation occurring at the joint interface (19). The former can be estimated from the change of density on solidification, but the latter may be difficult to compute since reorientation effects could originate through a crystallization process starting from some nucleus not located in the interface. Internal stresses and stress concentrations usually develop on solidification of the adhesive, the most common cause being the difference in the thermal expansion coefficients of the adhesive and adherend. In many applications of adhesives this matching process is not readily done or is neglected because it is not critical. In general, however, the theoretical strength of the adhesive joint is considerably decreased by the development of internal stress concentrations.

The effects of internal stresses developed in solidification and of stress concentrations during applications of a joint have been given much attention. An excellent discussion of this area of research has been given by Sneddon (74). Of particular interest here are the photoelastic studies of Mylonas (75) and Mylonas and deBruyne (76), which led to the conclusion that in a lap joint, poor wetting of the adherend tends to produce a greater stress concentration at the free surface of the adhesive where failure is most likely to be initiated. As the contact angles become large, the maximum stress concentration increases and moves toward the lineal boundary where the adhesive and atmosphere make contact with the adherend, the stress concentra-

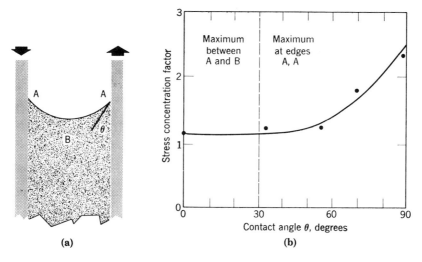

Fig. 10. Maximum stress concentration in a lap joint as a function of the degree of wetting.

tion factor increasing from about 1.2 to 2.5 (Fig. 10). Furthermore, Griffith (77) has shown that failure of the adhesive may occur at a relatively small applied stress if there are air bubbles, voids, inclusions, or surface defects; it occurs because stress concentrations result that are much higher than the mean stress applied across the specimen. This conclusion is especially important when considered in terms of the probable effect of poor wetting on the development of air pockets at the adhesive–adherend interface. It has already been shown that if $\theta = 0$, the theoretical joint strength far exceeds the tensile strength of most adhesives. In practice, the theoretical joint strength is not attained; evidently, one major cause is the development of stress concentrations during solidification of the adhesive. When $\theta = 0$, there may be gas pockets formed at the adhesive–adherend interface around which stress concentrations can build up, for if the adhesive is too viscous when applied, it may never penetrate the accessible surface pores before polymerizing. Of course, this situation is aggravated if $\theta \neq 0$.

In adhesives technology it is common practice to roughen the surface of each adherend, or "give it tooth," and so obtain a stronger joint. This practice can be theoretically justified with reservations by the following considerations. If gas pockets or voids in the surface depressions of the adherend are all nearly in the same plane and are not far apart (as on the upper adherend of Fig. 2), there may be crack propagation from one pocket to the next, and the joint may break as if it had a built-in "zipper." Therefore, if roughness must be accepted, the kind of roughness shown on the lower adherend would be preferable because crack propagation along a plane would be less probable (20).

Mechanism of Adhesive Action

An adequate theory of adhesive action should be able to explain poor as well as good adhesion; hence, a test of any theory is its ability to explain the properties of abhesives. *Abhesives* are films or coatings which are applied to one solid to prevent— or greatly decrease—the adhesion to another solid when brought into intimate contact with it.   Abhesives are employed in molding, casting, or rolling operations; therefore, it is common to refer to the film as the parting, mold-release, or antistick agent (see RELEASE AGENTS).   Examples of materials commonly used for such purposes are the polymethylsiloxanes; the high-molecular-weight fatty acids, amines, amides, and alcohols; various types of highly fluorinated fatty acids and alcohols; and Teflon or F.E.P. Teflon fluorocarbon films deposited by coating the mold with an aqueous dispersion of polymer particles, drying it, and finishing with a brief bake at a high temperature.   Usually a condensed monolayer of the agent is sufficient to cause the optimum parting effect.   Table 4 gives the values of γ_c of each of these films.   Evi-

Table 4. Critical Surface Tensions of Wetting of Surfaces Coated with Abhesives[a]

Coating material, all condensed films	γ_c, dyn/cm at 20°C
polymethylsiloxane film	24
fatty acid monolayer	24
polytetrafluoroethylene film	18
$HCF_2(CF_2)_n COOH$ monolayer	15
poly(1,1-dihydroperfluorooctyl methacrylate)	10
perfluorolauric acid monolayer	6

[a] From Ref. 19.

dently, each adhesive coating converts the solid into a low-energy surface, and any liquid placed on such a coated surface will exhibit an equilibrium contact angle which will be larger as $\gamma_{LV°} - \gamma_c$ increases.   When θ is large enough, such poor adhesion results that the application of a modest external stress suffices for effective mold release.   However, the excellent and easy parting action observed needs additional explanation.

It was pointed out earlier that the value of W_A for the various low-energy surfaces reported here exhibits a maximum variation of threefold in going from the least to the most adhesive materials.   This range is much too small to explain the effectiveness and easy parting action caused by a film of a good abhesive.   It is to be noted that the value of W_A is for a flat, nonporous, smooth surface.   The roughness factor r of the uncoated surface of the mold could suffice to raise W_A by a factor of from 1.5 to 3 or more, depending on the surface finish; however, if the material to be molded is (or becomes) viscous rapidly during injection or application, poor wetting will cause pockets to be produced at the interface between the material molded and the abhesive, and thereby the adhesion will be greatly decreased through stress concentrations by some unknown fraction $1/g$.   Also, when $\theta > 30°$, the resulting stress concentration factor of from 1.2 to 2.5 at the abhesive–plastic interface acts to decrease adhesion accordingly.   Therefore, the work of adhesion per unit area of the apparent or envelope area at the molding interface will be $(r/gs)W_A$.   For the best adhesive action, the release agent should have the largest possible value of θ for the substances to be molded against it.   The large values of θ encountered with the various types of organic liquids on such low-energy surfaces as those shown in Table 4 make it evident why these film-forming materials are so effective as abhesives.   In general, any low-energy surface will be more effective as a release agent (or abhesive) the lower its value of γ_c or the lower W_A (19).   It also can be concluded from the preceding discussion that the smoother the finish of the outer coating of the abhesive, and the lower the surface tension and viscosity of the material being molded (or the lower its contact angle with the adhesive), the greater will be the external stress required to cause the parting action.

Frequently it is desired that the release coating should be an integral part of the plastic object. An effective approach can be developed by including in the plastic, while it still is in the liquid state, a small concentration of an additive agent that is surface-active in the liquid and so is able to accumulate to some extent as an adsorbed film. The molecular structure of the additive agent should be such as to have the proper organophobic–organophilic balance with both types of groups located at opposite extremes of the molecule. Suitable organophobic groups would be those which have a perfluorinated chain, polymethylsiloxane structure, or a paraffin chain (Table 4) (78).

On the Adhesion Rule

In 1939 deBruyne (79) proposed the following rule for adhesives: "Provided we use pure or simple substances as adhesives then there is a good deal of evidence that strong joints can never be made by polar adherends with polar adhesives." This rule may be explained as follows (19): The statement that polar adhesives do not form strong joints with nonpolar adherends results of necessity from the fact that a liquid polar adhesive usually has a higher surface tension $\gamma_{LV^{\circ}}$ than the critical surface tension γ_c of wetting of a nonpolar adherend, and hence poor wetting would be encountered. There would result the previously mentioned difficulties with gas-pocket formation during spreading, stress concentrations at the joint if θ becomes large, etc. But a nonpolar adhesive liquid would usually have a lower surface tension $\gamma_{LV^{\circ}}$ than the critical surface tension γ_c of wetting, and spreading should result. However, many nonpolar adhesives are very hydrophobic materials, whereas many polar adherends are somewhat hydrophilic and so can absorb some water. Hence, in a somewhat humid atmosphere, poor wetting and spreading of the nonpolar adhesive would occur on contact with the slightly moist surface of the polar adherend. Many polar liquid adhesives, on the other hand, will be able either to absorb the atmospheric water or to displace through surface-chemical action the film of water adsorbed on the surface of the polar adherend; hence, adequate wetting and spreading of the adhesive may occur under ordinary atmospheric conditions. This mechanism can be enhanced sometimes by appropriate additives. It is therefore concluded that the deBruyne adhesion rule, although a useful guide, is not always correct and should be used with caution.

Since wetting is determined by the surface constitution of the adherend, it is evident that if one wishes to decrease joint strength one can modify the limitations implied in the deBruyne rule by absorbing on the surface of the adherend a monolayer of a parting agent or abhesive. If it is desired to increase joint strength, the surface-chemical composition of the adherend can often be chemically modified to increase γ_c. An example is the now widely used surface oxidation of polyethylene with chromic acid, flaming, ultraviolet radiation, or a brush discharge. Such treatments cause surface oxidation of the polyethylene, which will always increase γ_c and the wettability and adhesion by all liquids. Many common adhesives, although liquid, have values of $\gamma_{LV^{\circ}}$ that exceed γ_c of polyethylene. Raising γ_c makes smaller the difference $\gamma_{LV^{\circ}} - \gamma_c$ and decreases the contact angle. This was confirmed by the large increase in joint strength found upon surface oxidation with chromic acid (14).

Some Practical Considerations

The preceding analysis of adhesion and adhesives makes it evident that for optimum joint strength, it is essential to keep the contact angle between the liquid adhesive and the adherends as small as possible in order to minimize the buildup of stress concentrations and to obtain good spreading. Obviously, the interface of each adherend must be kept as smooth and free as possible of low-energy surface films and dust in order to prevent the formation of gas pockets and occlusions. In applying liquid adhesives the viscosity should be as low as possible in order to increase the extent of capillary flow into pores and crevices. Maximum spreading and capillarity will be obtained with adhesives having the highest surface tension compatible with obtaining a low contact angle. When conditions of complete wetting and freedom from the formation of gas pockets and occlusions prevail, the adhesion to either high-energy or low-energy surfaces will usually be ample, and generally failures of the joint will be in cohesion. If the surface of the adherend is to be roughened, it must be done so as to keep any surface gas pockets formed from being coplanar or nearly so over a large region. Under highly idealized conditions the problem of obtaining optimum joint strength with many adhesives and adherends does not appear to involve surface-chemical problems. However, the problem of coping with the rough surfaces encountered under practical conditions (and also under any conditions with certain indicated materials and surfaces) emphatically involves surface-chemical problems.

Because of the existence of pores and cracks in the surfaces of real solids, surface occlusions or gas pockets will always be formed to some extent on applying the adhesive. The resulting loss in joint strength can be large, and it may even nullify the theoretical gain in joint strength of using an adhesive with the surface tension given by equation 20, that is, such as to give the maximum value to W_A. It is possible that the strongest joints attainable in practice will be found when γ_{LV} of the adhesive is slightly less than γ_c of the adherend. In other words, more may be gained by minimizing the tendency to form interfacial occlusions than by maximizing the specific adhesive energy W_{A*}.

There has been much discussion about the necessity for using adhesives capable of forming chemical bonds with the adherends. But the preceding considerations concerning the relation of wetting to adhesion have made it evident that the energy involved in the physical adsorption to the adherend of molecules of adhesive is more than sufficient to form joints which are stronger than the solidified bulk adhesive. Of course chemical bonding may be desirable to obtain greater heat, water, or chemical resistance rather than any needed increase in the specific adhesion. For example, in making glass fiber–resin laminates, the use of coupling agents (or glass-fiber finishes) and resins with reactive terminal groups may have been evolved in order to react molecules of the agent with many accessible sites on the glass surface and so decrease its vulnerability to loss of strength by reaction with adsorbed water. However, it is usually a great advantage in selecting an adhesive for a given application to be released from the very limiting requirement of finding a material capable of chemically bonding to the adherends.

Future Research

Despite the advances in research on adhesion, much is yet to be learned. The treatment given here is based upon the equilibrium thermodynamics of the adhesive joint system; hence dynamic conditions of loading and joint fracturing have usually been ignored. As recently pointed out (8,70), attention needs to be given to the conditions of plastic flow occurring in many systems prior to the initiation of fracturing. The effects on joint failure of imperfections in the bulk of the adhesive have not been discussed here because these involve properties of the solid state that affect cohesive rather than adhesive properties of the joint.

Practically no reliable data are available on the effect of temperatures on the equilibrium contact angle. Several general effects of variations in the temperature and humidity on the contact angle can be expected. It is well known that the surface tensions (γ_{LV°) of pure organic liquids decrease linearly with rising temperature until close to the critical temperatures. Hence, if the temperature is T, then

$$\gamma_{LV^\circ} = C_1 - C_2 T \tag{30}$$

But since $\cos \theta = a - b\gamma_{LV^\circ}$, when the surface composition of the solid is constant, C_1, C_2, a, and b are each positive. Therefore

$$\cos \theta = (a - bC_1) + bC_2 T \tag{31}$$

and (assuming that a and b are not too dependent on temperature) $\cos \theta$ must increase (or θ decrease) with increasing temperature. In addition, the effect of raising the temperature will be to cause increased desorption of any physically adsorbed compounds. Temperature increase will therefore decrease the packing of such adsorbed films, and this will raise the critical surface tension of the system and thus cause θ to decrease.

However, when the temperature becomes high, there may result chemical changes in the liquids, such as hydrolysis, oxidation, and pyrolysis, or there may develop surface chemical changes in the solid due to oxidation, dehydration, or crystallographic rearrangement. The products of chemical reaction in the liquids may be highly adsorbable and may cause the formation of new low-energy surface films on which the liquids will not spread. This effect may overbalance the above-mentioned normal decrease in θ with rising temperature. The oxidation or dehydration of these inorganic solid surfaces may greatly alter the wetting behavior of the liquids. However, if the chemical reactivity or adsorptivity of the liquid is little altered thereby, no large change in wettability is to be expected.

Decreasing the relative humidity at ordinary temperatures will have little effect on θ unless the atmosphere is made so dry as to dehydrate the solid surface (as might occur in silica and α-alumina). But unusually long exposure to very dry air will be required to change significantly the contact angle. As the relative humidity approaches 100%, increased condensation of water on the surfaces of both metals and metallic oxides will invite the hydrolysis in situ of some adsorbed liquids, and so the wetting properties of these surfaces will become more like those of such highly hydrated surfaces as glass, silica, and alumina.

Obviously, data are needed on the values of f_{SV° for a variety of low-energy solid surfaces; with these and the values of W_{A*} given here, one would no longer have to estimate W_A, and hence our knowledge of adhesion would become more precise. It is also obvious that there is need for more reliable data on the surface tensions of liquid polymers, especially those in use, or potentially useful, as adhesives. In addition, the effect of solvents used as thinners is important insofar as: (a) they affect the surface tension of the polymer, and (b) if any polar adsorbable impurities are present they affect the wetting of high-energy surfaces.

The empirical nature of γ_c is obvious, and it would be helpful to replace γ_c by parameters having a sound basis in thermodynamic or statistical mechanical considerations. A simple consideration of the Young equation and the definition of γ_c indicates that when $\cos \theta = 1$, $\gamma_{SV^\circ} - \gamma_{SL} = \gamma_c$. Unfortunately, the effect of constitution on either γ_{SV° or γ_{SL} is still unknown, and neither quantity can be studied until a satisfactory experimental method for its measurement has been found. However, neither quantity appears to be a simple function of the constitution of the solid and liquid phases. A more precise theory of wetting of low-energy surfaces should at least include γ_{SL}. Probably the lateral spread in the data points of our graphs of $\cos \theta$ vs γ_{LV} for a given solid surface is due to the variation of γ_{SL} among the liquids used. Recent efforts by Fowkes (80,81) to relate γ_c to the dispersion forces between molecules of the contacting interfaces have been especially promising in leading to tractable equations. An interesting direct correlation has been recently pointed out by Gardon (82) between the value of γ_c of a solid polymer and the Hildebrand solubility parameter, δ, which is defined as the square root of the molar energy density (ie, $\delta = \sqrt{E/\rho}$). See COHESIVE-ENERGY DENSITY.

Although in the past adhesion has been an art, the basic principles and contributing mechanisms have become clearly defined; accelerating progress in developing and applying the subject can be expected in the next decade.

Bibliography

1. N. A. deBruyne and R. Houwink, eds., *Adhesion and Adhesives*, Elsevier Publishing Co., Amsterdam, 1951.
2. D. D. Eley, ed., *Adhesion*, Oxford Press, London, 1961.
3. P. Weiss, "Adhesion and Cohesion," *Proceedings of Symposium, General Motors Research Lab., Warren, Mich., July 1961*.
4. R. Holm, *Die Technische Physik der Electrischen Kontacte*, Springer-Verlag, Berlin, 1941.
5. F. P. Bowden and D. Tabor, *Friction and Lubrication of Solids*, Oxford Press, London, 1950.
6. O. L. Anderson, *Wear* **3**, 253 (1960).
7. J. S. McFarlane and D. Tabor, *Proc. Roy. Soc. (London) Ser. A* **202**, 224 (1950).
8. D. Tabor in D. D. Eley, ed., *Adhesion*, Oxford Press, London, 1961, Chap. 5.
9. R. C. Bowers, W. C. Clinton, and W. A. Zisman, *Mod. Plastics* **31**, 131 (1954).
9a. J. H. Poynting and J. J. Thomson, *Properties of Matter*, 13th ed., Charles Griffin, London, 1934, p. 152.
10. H. M. Budgett, *Proc. Roy. Soc. (London) Ser. A* **86**, 25 (1911).
11. Sir W. B. Hardy, *Collected Scientific Papers*, Cambridge Press, London, 1936.
12. S. H. Bastow and F. P. Bowden, *Proc. Roy. Soc. (London) Ser. A* **134**, 404 (1931).
13. N. A. deBruyne, *Research (London)* **6**, 362 (1953).
14. N. A. deBruyne, *Nature* **180**, 262 (Aug. 10, 1957).
15. J. W. McBain and D. G. Hopkins, *J. Phys. Chem.* **29**, 88 (1925).
16. J. W. McBain and W. B. Lee, *Ind. Eng. Chem.* **19**, 1005 (1927).
17. A. W. Hull and E. E. Burger, *Physics* **5**, 384 (1934).
18. W. A. Zisman, "Relation of Chemical Constitution to the Wetting and Spreading of Liquids on Solids," *A Decade of Basic and Applied Science in the Navy*, U.S. Govt. Printing Office, Washington, D.C., 1957, p. 30.
19. W. A. Zisman, "Constitutional Effects on Adhesion and Abhesion," in P. Weiss, ed., *Adhesion and Cohesion*, Elsevier Publishing Co., New York, 1962.
20. W. A. Zisman, "Relation of the Equilibrium Contact Angle to Liquid and Solid Constitution," Kendall Award Address, in *Contact Angle, Wettability, and Adhesion*, No. 43 in *Advances in Chemistry* Series, American Chemical Society, Washington, D.C., 1964, p. 1.
21. T. Young, *Phil. Trans. Roy. Soc. (London)* **95**, 65 (1805).

22. A. Ferguson, *Proc. Phys. Soc. (London)* **53**, 554 (1941).
23. H. W. Fox and W. A. Zisman, *J. Colloid Sci.* **5**, 514 (1950).
24. *Ibid.*, **7**, 109 (1952).
25. *Ibid.*, **7**, 428 (1952).
26. R. N. Wenzel, *Ind. Eng. Chem.* **28**, 988 (1936).
27. A. B. D. Cassie and S. Baxter, *Trans. Faraday Soc.* **40**, 546 (1944).
28. R. Shuttleworth and G. L. Bailey, *Discussions Faraday Soc.* **3**, 16 (1948).
29. C. G. Sumner, *Symposium on Detergency*, Chemical Publishing Co., New York, 1937, p. 15.
30. R. E. Johnson, *J. Phys. Chem.* **63**, 1655 (1959).
31. A. Dupré, *Théorie Mécanique de la Chaleur*, Gauthier-Villars, Paris, 1869, p. 369.
32. D. H. Bangham, *Trans. Faraday Soc.* **33**, 805 (1937).
33. D. H. Bangham and R. I. Razouk, *Trans. Faraday Soc.* **33**, 1459 (1937).
34. W. A. Cooper and W. H. Nuttall, *J. Agr. Sci.* **7**, 219 (1915).
35. W. D. Harkins and E. H. Grafton, *J. Am. Chem. Soc.* **42**, 2534 (1920).
36. W. D. Harkins and W. W. Ewing, *J. Am. Chem. Soc.* **42**, 2539 (1920).
37. W. D. Harkins and A. Feldman, *J. Am. Chem. Soc.* **44**, 2665 (1922).
38. W. D. Harkins, *Chem. Rev.* **29**, 408 (1941).
39. S. M. Skinner, R. L. Savage, and J. E. Rutzler, *J. Appl. Phys.* **24**, 438 (1953).
40. *Ibid.*, **25**, 1055 (1954).
41. S. M. Skinner, J. Gaynor, and G. W. Sohl, *Mod. Plastics* **33** (Feb. 1956).
42. B. V. Deryagin, *Research (London)* **8**, 70 (1955).
43. S. S. Voyutskii, *Adhesives Age* **5**, 30 (1962).
44. S. S. Voyutskii and V. L. Vakula, *J. Appl. Polymer Sci.* **7**, 475 (1963).
45. H. W. Fox, E. F. Hare, and W. A. Zisman, *J. Colloid Sci.* **8**, 194 (1953).
46. E. G. Shafrin and W. A. Zisman, *J. Phys. Chem.* **64**, 519 (1960).
47. H. W. Fox, E. F. Hare, and W. A. Zisman, *J. Phys. Chem.* **59**, 1097 (1955).
48. E. F. Hare and W. A. Zisman, *J. Phys. Chem.* **59**, 335 (1955).
49. M. K. Bernett and W. A. Zisman, *J. Phys. Chem.* **64**, 1292 (1960).
50. *Ibid.*, **65**, 2266 (1961).
51. E. F. Hare, E. G. Shafrin, and W. A. Zisman, *J. Phys. Chem.* **58**, 236 (1954).
52. F. Schulman and W. A. Zisman, *J. Colloid Sci.* **7**, 465 (1952).
53. A. H. Ellison, H. W. Fox, and W. A. Zisman, *J. Phys. Chem.* **57**, 622 (1953).
54. *Ibid.*, **58**, 260 (1954).
55. *Ibid.*, **58**, 503 (1954).
56. F. London, *Trans. Faraday Soc.* **33**, 8 (1937).
57. J. O. Hirschfelder, C. F. Curtiss, and R. B. Bird, *Molecular Theory of Gases and Liquids*, John Wiley & Sons, Inc., New York, 1954.
58. E. G. Shafrin and W. A. Zisman, *J. Phys. Chem.* **61**, 1046 (1957).
59. *Ibid.*, **66**, 740 (1962).
60. I. Langmuir, *J. Am. Chem. Soc.* **38**, 2286 (1916).
61. I. Langmuir, *Trans. Faraday Soc.* **15** (3), 62 (1920).
62. I. Langmuir, *Third Colloid Symposium Monograph*, Chemical Catalogue Co., New York, 1925, p. 48.
63. I. Langmuir, *J. Franklin Inst.* **218**, 143 (1934).
64. I. Langmuir, *Collected Works*, Pergamon Press, New York, 1961.
65. P. R. Basford, W. D. Harkins, and S. B. Twiss, *J. Phys. Chem.* **58**, 307 (1954).
66. W. C. Bigelow, E. Glass, and W. A. Zisman, *J. Colloid Sci.* **2**, 563 (1947).
67. G. E. Boyd and H. K. Livingston, *J. Am. Chem. Soc.* **64**, 2383 (1942).
68. W. D. Harkins and E. H. Loeser, *J. Chem. Phys.* **18**, 556 (1950).
69. E. H. Loeser, W. D. Harkins, and S. B. Twiss, *J. Phys. Chem.* **57**, 251 (1953).
70. D. Tabor, *Rept. Prog. Appl. Chem.* **36**, 621 (1951).
70a. G. Kraus and J. E. Manson, *J. Polymer Sci.* **6** (5), 625 (1951).
71. K. W. Bewig and W. A. Zisman, "Low Energy Reference Electrodes for Investigating Adsorption by Contact Potential Measurements," in *Solid Surfaces and the Gas-Solid Interface*, No. 33 in *Advances in Chemistry* Series, American Chemical Society, Washington, D.C., 1961, p. 100.
72. J. M. Martinet, "Adsorption des Composés Organiques Volatiles par le Polytétrafluor Éthylene," *Commissariat à l'Énergie Atomique, Rapport CEA* **888**, Centre d'Études Nucléaires de Socloy, 1958.

73. D. Graham, *J. Phys. Chem.* **66**, 1815 (1962).
74. I. N. Sneddon, "The Distribution of Stress in Adhesive Joints," in D. D. Eley, ed., *Adhesion*, Oxford Press, London, 1961, Chap. 9.
75. C. Mylonas, *Proc. Intern. Congr. Appl. Mech., 7th, London,* **1948**.
76. C. Mylonas and N. A. deBruyne, in N. A. deBruyne and R. Houwink, eds., *Adhesion and Adhesives,* Elsevier Publishing Co., Amsterdam, 1951, Chap. 4.
77. A. A. Griffith, *Phil. Trans. Roy. Soc. (London) Ser. A* **221**, 163 (1920).
78. N. L. Jarvis, R. B. Fox, and W. A. Zisman, "Surface Activity at Organic Liquid–Air Interfaces. V. Effect of Partially Fluorinated Additives on Wettability of Solid Polymers," in *Contact Angle, Wettability, and Adhesion,* No. 43 in *Advances in Chemistry* Series, American Chemical Society, Washington, D.C., 1964.
79. N. A. deBruyne, *The Aircraft Engineer* **18** (12), 53 (Dec. 28, 1939).
80. F. M. Fowkes, *J. Phys. Chem.* **66**, 382 (1962).
81. F. M. Fowkes, "The Dispersion Force Contributions to Surface and Interfacial Tensions, Contact Angles, and Heats of Immersion," in *Contact Angle, Wettability, and Adhesion,* No. 43 in *Advances in Chemistry* Series, American Chemical Society, Washington, D.C., 1964.
82. J. L. Gardon, "Relationship Between Cohesive Energy Densities of Polymers and Zisman's Critical Surface Tensions," *J. Phys. Chem.* **67**, 1935 (1963).

ACKNOWLEDGMENT. The author thanks the American Chemical Society for permission to use portions of his paper, *Ind. Eng. Chem.* **55** (10), 18–38 (1963).

W. A. Zisman
U.S. Naval Research Laboratory

THEORY OF ADHESIVE JOINTS

This section deals with the theoretical basis for the strength of adhesive joints, and more particularly with the resistance to rupture of a bonded assembly. See also Design of Adhesive Joints under Bonding, p. 522.

Bonded assemblies or adhesive joints are comprised of *adherends* (ie, bodies held together by the adhesive) and the *adhesive film*, although weak boundary layers (see below) may be more important than the visible components. Both the adherends and the adhesive may be polymeric. The strength of a joint has only a very loose connection with molecular adhesion, because when a joint is broken by mechanical means, rupture practically never proceeds along the adherend–adhesive boundary (1); thus, forces across the boundary do not influence the measurable force needed for breaking the joint. The reasons that the rupture surface does not coincide with the adherend–adhesive interface are that in many instances such an interface does not exist (ie, adherend and adhesive get intermixed during the preparation or aging of the joint), and if it exists, separation along an exactly predetermined, complicated path is extremely improbable.

The history of an adhesive joint usually comprises three periods. In the first, the adhesive is applied to the adherends and the joint is formed; in the second, the adhesive sets (ie, solidifies); and in the third, the properties of the joint remain sensibly constant (this is the final state). Many joints, in addition to the adherend and the adhesive generally mentioned when the joint is described, also contain a third phase spread along the adherend–adhesive boundary. This phase is known as a *boundary layer*. What happens to the three components of the joint during the three periods of its history depends on the type to which the joint belongs: hooking, improper, or proper.

Types of Joints

Hooking Joints. These joints are common with fibrous adherends such as paper, paperboard, or fabrics. When the adhesive is applied (as a liquid), it penetrates into the pores or voids between the fibers and, after setting, forms hooks or loops around these fibers; thus, *mechanical interlocking* ensues. In the simplest instance, the adhesive is a Newtonian liquid of viscosity η, and the pores are slits of width δ. The rate of capillary penetration is then given by equation 1, in which z is the depth

$$\frac{z}{t} = \frac{\delta \, \gamma \cos \theta_A}{3\eta z} \tag{1}$$

35

of penetration reached after time t, γ is the surface tension of the liquid, and θ_A is the advancing contact angle (2). Industry as a rule favors rapid formation of joints, so the adhesive liquid should not be too viscous and should completely wet the adherend (ie, θ_A should equal zero). After solidification, the final state is reached, and the breaking stress of the joint in this state depends on the value of z reached during the adhesive application (3).

Mechanical interlocking usually is unimportant when the adherends are not porous or at least have a nonporous surface. However, an equation analogous to equation 1 is still valid during the formation stage. All solid surfaces are rough; the adhesive has to fill the valleys on these surfaces, and z in these instances means the depth of penetration of the liquid front into a valley. As valleys are rarely slitlike, equation 1 cannot be expected to give quantitative information. It may be stated, however, that satisfactory (or practically complete) filling of the valleys will take place if the dimensionless ratio $\gamma \cos \theta_A t/\eta z$ is greater than a number characteristic for any given surface; for a surface carrying slitlike depressions this number is $3z/\delta$.

If the valleys are filled with the liquid adhesive and the two adherends are separated before setting takes place, the resistance to separation determines the *tack* (or *tackiness*) of the adhesive. If the liquid adhesive fills the space between two parallel adherend discs of radius a, and if it is also present outside this space, then equation 2 holds approximately (4).

$$ft = 0.75 \, \eta \, \frac{a^2}{h^2} \tag{2}$$

In this equation, f is the stress causing separation in t seconds, and h is the initial distance between the discs. If there is no adhesive outside the narrow space between the discs, then equation 3 holds.

$$\left(f - \frac{2\gamma}{h}\right) t = 0.75 \, \eta \, \frac{a^2}{h^2} \tag{3}$$

Improper Joints. Improper joints contain a weak boundary layer in which rupture takes place when an external force is applied to the joint. Proper joints do not contain this layer. (In proper joints, as in hooking joints, either the adhesive or the adherend breaks.) The weak boundary layers may be of many kinds and can form during the formation or during the setting periods.

A convenient classification of weak boundary layers is based on the origin of the material making up the layer. This material may originate in the surroundings (usually air), in the adhesive, in the adherend, or from two sources together; air and adhesive, air and adherend, adhesive and adherend, or even all three. A weak boundary layer of air pockets forms during the application of the liquid if the valleys on the adherend surface are not filled with the adhesive. This is the reason for the well-known rule that the adhesive must *wet* the adherends well if a strong bond is desired.

If a material contains substances that have a low tensile or shear strength, are soluble in the main component as long as this is liquid, and are insoluble in the solidified main component, then this material will give rise to weak boundary layers when it is present as adherend or used as an adhesive. Commercial polyethylene was best studied in this respect. Many industrial polyethylenes, when molten on a solid support and then permitted to solidify, can be pried off the support with a fingernail.

No printing ink or other adhesive matter adheres to these samples. However, if the impurities causing formation of weak boundary layers are removed (eg, by fractional precipitation), the purified polyethylenes are as capable of adhesion as any other polymer. If controlled amounts of impurities, such as oleic acid (5,6), are introduced into a purified polyethylene, its ability to adhere is reduced to almost nothing when the impurity concentration exceeds the solubility limit in the host substance (ie, in the purified polyethylene).

A weak boundary layer originating in three phases (7) was observed when a complex polymeric mixture was cured in air between two copper adherends. Copper accelerated oxidation of the polymer by air, and to avoid improper joints it was sufficient to substitute aluminum or some other metal for the copper, to displace the air by nitrogen, or to change the composition of the adhesive.

If a polymer "does not stick to anything," very probably it contains impurities, such as lubricants, antioxidants, and so on, which act analogously to oleic acid in purified polyethylene. If a polymer readily adheres to many solids but does not adhere to a few others, the latter perhaps form an oily compound with the polymer, or behave analogously to copper in the example cited above.

Proper Joints. For the strength of proper joints in their final state the approximate equation

$$f = \frac{1}{\alpha}\left(\frac{\xi}{\beta} - s\right) \tag{4}$$

may be used. In it, f is the breaking stress, ξ is the cohesion of the adhesive material (if it is a near-Hookean solid and if it breaks rather than the adherend), α is the stress concentration factor due to the difference between the mechanical constants of the adherend and the adhesive β is the analogous factor due to the heterogeneity of all solids, and s is the shrinkage or frozen stress.

When the chemical composition of an adhesive is varied, the primary effect is on the value of ξ, but this effect is usually moderate. Factor β, on the other hand, may vary within a wide range (for example, 10–1000) depending on the conditions of solidification of the adhesive. Thus, if ξ is raised by a chemical change in the adhesive but at the same time the formability of the adhesive material is reduced (making β greater), the overall effect easily can be to lower the ratio ξ/β and with it the value of f. As a crude approximation, ξ/β is the tensile strength (or shear strength) of the adhesive substance.

The meaning of factor α will be clear from Figure 1. Figure 1a is a butt joint not deformed by an external force; Figure 1b is the same joint after deformation. The adherends are supposed to be rigid, and the adhesive is rubbery. It is seen that the force, although it acts in an up-and-down direction in the figure, extends the adhesive film also in the right-and-left direction near the adherend boundary. Calculation and photoelastic measurements (11) demonstrate that greatest stress concentrations occur near points aa' and bb' of the joint; α of equation 4 is the ratio of the maximum stress at one of these points to the average stress in the adhesive film. The value of α is smaller, ie, f is greater, as the difference between the moduli of elasticity of adherend and adhesive becomes smaller; of course, lowering α should not lower the ratio ξ/β or increase s. The shrinking stress s, which commonly depresses f, can be reduced either by using adhesives that do not greatly contract during solidification or those

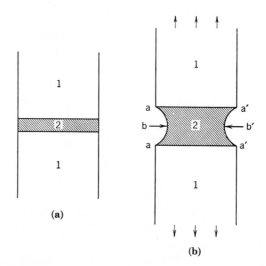

Fig. 1. A butt joint. (a) Before extension. (b) During extension.   1, adherend; 2, adhesive.

that can relax during contraction. In the latter case, although the strain caused by setting is large, the corresponding stress is small; adhesives in which this relaxation occurs may be used as sealers.

Whereas equation 4 is supposed to represent the behavior of joints that break without large deformation, equation 5 has been derived (8) for rigid-plastic materials;

$$f = \frac{2ka}{3h} \tag{5}$$

k is the lowest shearing stress at which plastic flow starts (yield value), a is the radius of the discs glued together (butt joint), and h is the initial distance between the discs. A comparison of equations 5 and 2 is instructive.

According to equation 5, the breaking stress of a butt joint should be inversely proportional to the thickness h of the adhesive film. Equation 4 contains the effect of h in a hidden form in the magnitudes α, β, and s. All three depend on h; it is clear from inspection of Figure 1b that stress distribution will be different at different thicknesses. The "waist" visible in Figure 1b cannot form in a very thin film; hence, extension in very thin films in the up-and-down direction causes an increase in volume. Extension of a thick film may entail no volume change, and for materials such as rubber this is a fundamental difference. The effect of h on β is such that β is smaller when h is smaller. This is one of the reasons for the often-observed increase of f when h decreases.

Equations 4 and 5 are applicable most readily to *butt joints*. In industry *lap joints* are usually preferred. An example is shown in Figure 2. The force F needed to split such a joint increases with the length l of the overlap; thus, by increasing l it is possible, as a rule, to make the joint stronger than the adherend bars so that the adherend breaks when an external force is applied. The dependence of F on l has been studied both theoretically and experimentally (1,9).

An illustration of a *peel joint* is given in Figure 3. If the dimensions of the adhesive film in such a system are length l, width w, and thickness h, then, in the simplest instance, the peeling force is proportional to wh, whereas the force needed to break a

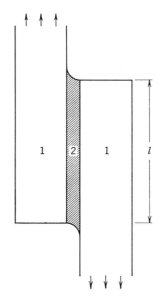

Fig. 2. A lap joint. 1, adherend; 2, adhesive; *l*, overlap

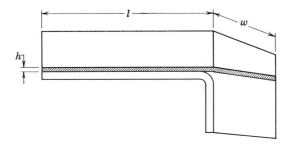

Fig. 3. A peel joint. Adhesive is shaded.

butt joint of identical dimensions would be proportional to *lw*. Thus, peeling is easier than any other form of rupturing a joint. Whenever peeling is possible, it tends to occur before another type of break.

Adhesives

A good adhesive must not contain substances capable of forming weak boundary layers. A polymer that can, in addition, dissolve or otherwise remove bond-weakening impurities from the surface of usual adherends is even better; this type can be used as a household adhesive that "sticks everything to everything." If the danger of weak boundary layers is avoided, the polymer that has a high value of the ratio ξ/β is generally good. In particular instances, the polymer should not cause the stress-concentration factor α to be too great, and should not give rise to excessive shrinkage stresses. For a rigid-plastic polymer, the yield value k should be large. These remarks presuppose that the mechanical strength of the joint in a noncorrosive environment is the most important property desired. If the joint has to be strong under water, or at high temperatures, or on impact, then additional considerations are called for.

The main technical reasons for using adhesives rather than other means of fastening are that no damage is done to the adherends (as in nailing), and that stress is more uniformly distributed (no stress concentration as at a nail or a bolt).

Because adhesive joints are used in very different structures subjected to very different conditions, tests simulating reality should be employed whenever possible. In addition, many "standard" tests are available. Often these tests are not sufficiently specific to enable one to predict the behavior in service, and not sufficiently scientific to permit one to base theoretical predictions on their results. A description of a group of such tests is available (10). For paper, tests T463 rm-52 and T806 sm-46 of the Technical Association of the Pulp and Paper Industry may be mentioned.

Bibliography

1. J. J. Bikerman, *The Science of Adhesive Joints*, Academic Press, Inc., New York, 1961, p. 125.
2. J. J. Bikerman, *Tappi* 44, 568 (1961).
3. J. J. Bikerman and W. Whitney, *Tappi* 46, 420, 689 (1963).
4. J. Stefan, *Sitzber. Akad. Wiss. Wien, Math.-Naturw. Kl.* 69, 713 (1874).
   For a discussion of this paper, see Ref. 1, p. 53.
5. J. J. Bikerman, *J. Appl. Chem.* 11, 81 (1961).
6. J. J. Bikerman and D. W. Marshall, *J. Appl. Polymer Sci.* 7, 1031 (1963).
7. J. M. Black and R. F. Blomquist, *Adhesives Age* 2 (5), 34 (1959); 2 (6), 27 (1959).
8. R. T. Shield, *Quart. Appl. Math.* 15, 139 (1957).
9. J. L. Lubkin and E. Reissner, *Trans. Am. Soc. Mech. Engrs.* 78, 1213 (1956).
10. *ASTM Standards on Adhesives*, 5th ed., American Society for Testing and Materials, Philadelphia, 1961.
11. Ref. 1, p. 145.

J. J. Bikerman
Massachusetts Institute of Technology

SURFACE PROPERTIES

A proper exposition of the surface properties of polymers with respect to adhesion and adhesive joint strength should begin by defining these terms. Adhesion, as used in this article, will refer only to the attractive forces exerted between a solid surface and a second phase (either liquid or solid). Adhesion is concerned with the phenomena involved in making an adhesive joint (ie, wettability, relative surface energetics of both phases, kinetics of wetting). These are purely surface considerations. Adhesive joint strength is the breaking strength of a bonded assembly. Once an adhesive joint has been formed, interfacial forces are no longer of primary concern, since interfacial failure probably never occurs under ordinary testing conditions, provided the surfaces were in intimate contact prior to breaking the adhesive joint. What is of prime importance is the mechanical response of the composite to an applied stress. Probably a more realistic description of the breaking strength of an adhesive joint would be based on a mechanical deformation theory of adhesive joint strength. See also ADHESION AND BONDING.

Formation of Polymer–Substrate Interface

Two materials probably adhere, at least initially, because of van der Waals attractive forces acting between the atoms in the two surfaces. Interfacial strengths, based on van der Waals forces alone, far exceed the real strengths of one or other of the adhering materials. Therefore, interfacial separation is unlikely when mechanical forces are used to separate a pair of materials which have achieved complete interfacial contact (probably a highly unlikely situation), or a number of separate regions of interfacial contact. Presumably, the breaking strength of a joint is not directly related to the nature of the interfacial forces acting between the adherends.

Van der Waals forces are operative over very small distances. Hence, in order that materials adhere, the atoms in the two surfaces must be brought close enough together for these forces to become operative. Adhesion, the joining of the two surfaces, is only a partial requirement for forming a strong adhesive joint. In addition to bringing the two surfaces together there must be an absence of weak boundary layers (surface regions of low mechanical strength), either present or generated by an interaction between both phases. The weak boundary layer concept will be treated in the next section.

If A (solid) and B (solid), each having an absolutely smooth (on an atomic scale) planar surface and no mechanical weakness in their surface regions, were joined in a perfect vacuum, all attempts to get them apart mechanically would result in failure in

41

either A or B (Fig. 1a). However, real surfaces differ from these ideal surfaces in that they are rough and contaminated. Both of these imperfections contribute to a greatly decreased real area of contact between the surfaces of A and B (Fig. 1b). In general,

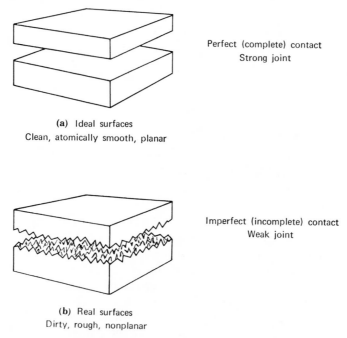

Perfect (complete) contact
Strong joint

(a) Ideal surfaces
Clean, atomically smooth, planar

Imperfect (incomplete) contact
Weak joint

(b) Real surfaces
Dirty, rough, nonplanar

Fig. 1. Surfaces: **(a)** ideal, and **(b)** real.

when interfacial contact has been established (ie, van der Waals forces are operative between A and B), they have adhered. When the joint is separated mechanically a little of A remains on B, or of B on A, depending on the geometry in the neighborhood of each area of contact and the mechanical strength in the surface region of A and B. The general reaction based on visual examination of a failed weak joint is that the solids did not adhere or the failure was at the interface. The first statement is incorrect, because surely some areas of A and B achieved interfacial contact; the second is incorrect because where the surfaces were not in interfacial contact prior to breaking the joint there was no adhesion.

Therefore, to form strong joints between A and B an increase in the real area of contact is needed as well as the absence of weak boundary layers. To achieve this, one or both of the materials to be joined must be made to conform better to the surface roughness of the other. This implies, in a practical sense, that one of the materials should be fluid when placed in contact with the other. It is necessary, but may not be a sufficient condition, that one of the materials be fluid. For example, if a high viscosity fluid makes a sizable contact angle with the solid (Fig. 2a), its tendency to create a large area of interfacial contact may be relatively small. The result is that it may do a great deal of bridging (ie, trap air and achieve little penetration into the surface roughness of the solid) and stress concentrations due to a large contact angle become important when the fluid solidifies. However, if the fluid member spontaneously spreads on the solid, $\gamma_{SV} - \gamma_{SL} \geq \gamma_{LV}$ (where γ is the surface tension and

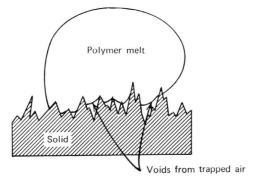

(a) Note low real area
of interfacial contact

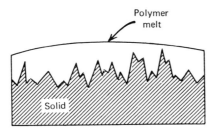

(b) Note lack of voids
from trapped air in
pores and crevices

Fig. 2. (a) Poorly wetted surface, high-viscosity fluid giving low real area of interfacial contact; (b) extensive intermolecular contact, low-viscosity fluid with absence of voids from trapped air in pores and crevices.

the subscripts refer to the phase), the interfacial area of contact increases, because the fluid can now flow more completely into the micro- or submicroscopic pores and crevices in the surface of the solid and can displace gas pockets and other contamination (Fig. 2b). In addition, the zero contact angle tends to minimize stress concentrations on solidification. The effect of creating a spontaneous spreading situation, then, is twofold: the real area of contact is increased and stress concentration is minimized.

Only van der Waals forces have been mentioned in the preceding treatment; this is not to be construed as meaning that other molecular forces may be excluded from participation in adhesion. In the initial process involving the establishment of interfacial contact, all those molecular forces involved in wetting phenomena can be considered to be important. Chemisorption is not excluded, but if it is to occur, molecular contact must have already been established, that is, van der Waals forces must already be operative (see also ADSORPTION; SORPTION). Therefore, any such chemical reaction as does occur, occurs after adhesion is complete. Further, since sensible interfacial separation does not occur under mechanical influences even when only van der Waals forces are operative, it follows that chemisorption can have no positive influence on the mechanical strength of an adhesive joint. It may have a negative influence on the strength of an adhesive joint if weak boundary layers are formed. However, it is possible that chemisorption may increase the permanence of an adhesive joint by retarding or preventing destruction of the interfacial region, as by moisture, low surface tension liquids, etc.

Studies of strong adhesive joints obtained in the epoxy adhesive–polyethylene and the epoxy adhesive–chlorotrifluoroethylene homopolymer, copolymer, and terpolymer systems clearly demonstrate the importance of the glass-transition temperature of the substrate, the surface roughness of the substrate, and the surface tension of the adhesive, provided no weak boundary layers are present or generated (1–3).

Weak Boundary Layers

Effect on Adhesive Joint Strength. Although the wettabilities of both chlorotrifluoroethylene homopolymer and polyethylene are similar ($\gamma_c \cong 31$ dyn/cm, where γ_c is the critical surface tension of the polymer, ie, that value of γ_{LV} at which a liquid has a contact angle of $0°$ with the polymer surface), strong joints can be made with the former but not with the latter at temperatures well below their respective melting temperatures. Recent work has indicated that polyethylene and other melt-crystallized polymers (eg, poly(vinyl fluoride), nylon-6,6 etc) have weak boundary layers in their surface regions which preclude the formation of strong adhesive joints even though extensive interfacial contact between the polymer film and the epoxy adhesive is achieved.

Apparently these weak boundary layers are generated in the process of preparing the polymer film. For example, strong joints cannot be made to untreated polyethylene film, yet polyethylene, when melted onto a metal adherend, results in strong adhesive joints when the metal is properly prepared and there is no mechanical weakness on the surface or in the metal oxide. The nature of the substrate surface has a profound effect on the surface properties of the films generated in contact with the substrate.

At a polymer melt–vapor interface or a solid–melt interface when the solid does not act as a nucleating site, species which cannot be accommodated into the crystal structure are rejected to the interface during the crystallization process, resulting in a surface region of low mechanical strength. Since many polymers are molded against non-nucleating surfaces, thereby generating surface regions of low mechanical strength, the elimination of these weak boundary layers becomes important.

Frictional Behavior. An experimental study of a wide range of polymeric solids shows that, as with metals, their friction is determined primarily by the shear strength of the junctions formed at the interface. With these materials, the area of contact depends upon their geometry as well as upon the load. Further, the deformation, or "ploughing," term in the friction expression arises not from plastic deformation of the solid but from elastic hysteresis losses. These factors are very dependent on time and temperature.

In a simple way, friction is thought to arise from two main causes. The first is the strong adhesive joint that may occur at the regions of real contact; junctions are formed which must be sheared when sliding takes place. The force to shear these junctions is S. The second mechanism assumes that if the asperities on the harder surface plough through the surface of the softer one, an extra ploughing force, P, is involved. The total observed friction, F, is then the sum of these:

$$F = S + P \tag{1}$$

For most situations $S \gg P$ and therefore F can be equaled with S. However, the ploughing force may constitute a large part of the friction, particularly if the interfacial adhesive joint strength is small. Apparently there is sufficient evidence to show that F is proportional to the load and independent of the area of the surfaces. Accordingly, the coefficient of friction, μ, is given by equation 2, where W is the load.

$$\mu = F/W \tag{2}$$

The coefficients of friction for a variety of polymers exhibiting weak boundary layers are presented in Table 1. Polymers having weak boundary layers invariably have low coefficients of friction. For example, polytetrafluoroethylene has long been regarded as a material with a low coefficient of friction, and it is also a material with a weak boundary layer. Surface treatment of the polymer to remove the weak boundary layer will increase the coefficient of friction. Removing the low-shear-strength surface layer increases the strength of the junctions formed between the polymer and stylus. Thus, providing a surface with increased mechanical strength results in an increase in

Table 1. Friction of Polymers (Metal on Polymer)

Polymer	Repeating unit	Coefficient of friction $(\mu)^a$
polyethylene	$-CH_2-$	0.6–0.8
polystyrene	$-CH_2-\overset{\displaystyle C_6H_5}{\underset{\displaystyle \vert}{CH}}-$	0.4–0.5
poly(vinyl chloride)	$-CH_2-CHCl-$	0.4–0.5
nylon-6,6	$-CO-(CH_2)_4-CO-NH-(CH_2)_6-NH-$	0.3
poly(methyl methacrylate)	$-CH_2-\overset{\displaystyle CH_3}{\underset{\displaystyle COOCH_3}{C}}-$	0.4–0.5
polytetrafluoro-ethylene	$-CF_2-CF_2-$	0.05–0.1

^{a} Dimensionless quantity.

both the shear strength of an adhesive joint and the coefficient of friction. See also ABRASION RESISTANCE.

Surface Treatments

The effect of weak boundary layers on adhesive joint strength may be reduced or eliminated by proper treatment of the polymer surface to produce either surface crosslinking or extensive nucleation at the surface during crystallization.

Surface Crosslinking. Conventional techniques for treating polyethylene surfaces in order to obtain strong adhesive joints, such as surface oxidation by corona discharge or flame treatment, are commonly believed to be effective because they create wettable polar surfaces onto which the adhesive may spread spontaneously and thus provide extensive interfacial contact. However, as pointed out previously, extensive interfacial contact is a necessary but not sufficient condition for forming strong joints. Apparently the primary function of surface oxidation techniques is to eliminate the weak boundary layer. If surface oxidation alone occurred without removal of the weak boundary layer, only weak adhesive joints would be obtained.

The low mechanical strength of the weak boundary layer, which prevents the formation of strong adhesive joints, can be increased rapidly and dramatically by allowing electronically excited species of rare gases to impinge upon the surface of the polymer. As these metastable and ionic gases come in contact with polyethylene, for example, they cause abstraction of hydrogen atoms and formation of polymer radicals at and near the surface of the polymer. The radicals formed by this process interact to form crosslinks and unsaturated groups without appreciable scission of the polymer chain. The mechanical strength of the surface region is increased remarkably by the formation of a gel matrix. Wettability of the surface is relatively unaffected. For example, the critical surface tension of wetting of polyethylene was approximately 35 dyn/cm both before and after treatment. This surface treatment technique is called casing (Crosslinking by Activated Species of INert Gases). Casing of polymers permits the formation of strong adhesive joints with conventional adhesives.

Contact time of activated gas with polymer film of as little as 1 sec under relatively mild conditions resulted in greatly improved adhesive joint strength for an epoxy adhesive on polyethylene. Longer contact times were required for polymers such as polytetrafluoroethylene. Excited helium, argon, krypton, neon, and xenon, and even hydrogen and nitrogen, are all effective crosslinking agents, although nitrogen changed the wettability of the surface.

The values obtained for tensile shear strengths of joints with polyethylene and polytetrafluoroethylene are shown in Figure 3. A tenfold or greater increase in lap shear joint strength was produced by bombardment with activated research grade helium although no change in wettability of the polymer was observed. The rise in joint strength for the *untreated* polyethylene lap shear composites results from a melting of the polymer onto the cured epoxy adhesive. Infrared examination (attenuated total reflectance) of polyethylene film which had been held at 0.05 mm pressure for several hours, then bombarded for 1 hr with activated research grade helium at 1 mm pressure, and finally kept at a pressure of 0.05 mm for 16 hr at room temperature to permit dissipation of radicals formed during bombardment, showed only the formation of transethylenic unsaturation in the surface region. Transmission spectra of treated and untreated films were identical, indicating that unsaturation, as was the case for crosslinking, occurs only at or near the surface of the polymer during bombardment. The spectra have no peaks attributable to carbonyl or hydroxyl groups. The improvement in adhesive joint strength achieved by treatment with inert gas is primarily due to the increase in the mechanical strength of the polymer in the surface region through formation of a densely crosslinked matrix.

Heterogeneous Nucleation. An important question is raised by the observation that the melting of a polymer onto a high-energy surface (ie, metal, metal oxide) generates a strong joint, provided the polymer has wet the substrate, whereas the free polymer film prepared by conventional techniques requires a surface treatment prior to joining. Apparently the substrate has a profound effect on the ultimate mechanical properties in the interfacial region of the polymer.

At the high-energy solid–polymer melt interface a region of substantial mechanical strength is generated by extensive wetting and subsequent nucleation and crystallization of the polymer. During crystallization from the melt, species contributing to the generation of weak boundary layers are rejected from the interface into the bulk. The resulting interfacial zone of high mechanical strength is illustrated in Figure 4 for the polyethylene–aluminum (chemically etched) system. Here a transcrystalline region is

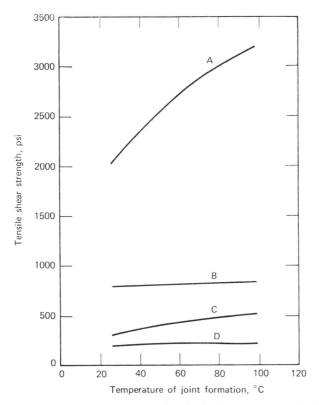

Fig. 3. Tensile shear strengths of aluminum–epoxy adhesive–polymer film–epoxy adhesive–aluminum composites as a function of the ultimate temperature of joint formation. A, glow-discharge treated in helium for 10 sec; B, glow-discharge treated in helium for 10 min; C, untreated polyethylene film; D, untreated polytetrafluoroethylene.

Fig. 4. Transcrystalline region in polyethylene formed in contact with etched aluminum foil. Foil dissolved from polymer prior to sectioning.

generated in the polymer surface region. Apparently, at the surface of the metal oxide, numerous crystallization nuclei are formed (Fig. 5a); the spherulites which grow from the nuclei now can propagate in only one principal direction, since growth in the lateral directions is inhibited by neighboring spherulites. In this way, only very narrowly divergent spherulite sectors develop, which give an overall appearance of rod-like buildup. The thickness of this transcrystalline region is estimated to be about

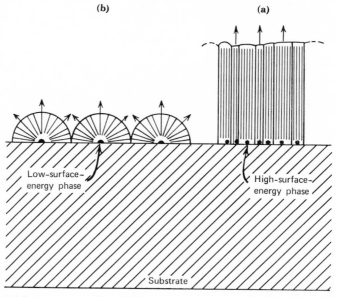

Fig. 5. Model for transcrystalline region. (**a**) Nucleation and crystallization of polymer melt at solid–liquid interface resulting in a transcrystalline region. (**b**) Bulk nucleation and crystallization, no transcrystalline region generated at the interface.

25–50 μm. There is no observed transcrystalline growth of the polyethylene surface when generated against a low-energy phase (Fig. 5b). As described earlier, poor nucleating phases generally contribute to the generation of weak boundary layers.

Solvent extraction (eg, xylene) of the transcrystalline region showed no evidence of a gel fraction. Apparently the increase in the mechanical strength of the surface region, as a result of the extensive nucleation, is due to considerable entanglement of the polymer chains. There is a strong competition for these chains since so many nuclei are formed. In effect, each entanglement may be considered to be a crosslink (Fig. 6).

Since strong adhesive joints can be formed by melting onto a high-energy surface, we can inquire whether the surface generated at the high-energy solid–polymer melt interface is amenable to conventional adhesive bonding when the metal is removed. This is indeed the case. As shown in Figure 7, it is important to remove the metal by dissolution rather than by peeling. Peeling disrupts the surface region of interest. This can be seen by examining the bondability of both the polymer and foil surfaces after peeling the foil from the polymer. In both cases the joint strengths are low (bottom curve, Fig. 7). When the foil is peeled, cohesive failure occurs in the polymer exposing two new surfaces which are not amenable to adhesive bonding.

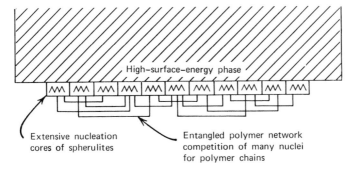

High–surface–energy phase

Extensive nucleation
cores of spherulites

Entangled polymer network
competition of many nuclei
for polymer chains

Fig. 6. Entanglements resulting from extensive nucleation and crystallization of polymer melt at solid–liquid interface.

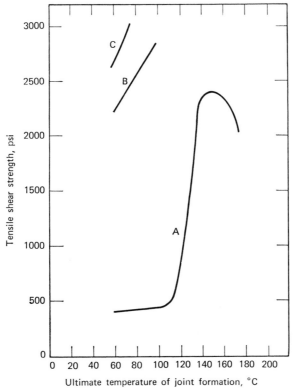

Ultimate temperature of joint formation, °C

Fig. 7. Tensile shear strength of aluminum–epoxy adhesive–polyethylene–epoxy adhesive–aluminum composites plotted as a function of the ultimate temperature of joint formation. A, Untreated film, below melting point, weak boundary layer. Above melting point, polymer nucleated on cured epoxy to preclude formation of weak boundary layer. B, Polyethylene nucleated and crystallized in contact with gold. Gold removed by amalgamating with mercury. C, Polyethylene treated in helium for 10 sec.

From the above analysis it can be concluded that the weakness in the surface region of many polymers, particularly polyethylene, is not an intrinsic property of the polymer but is dependent on the manner in which the surface region is formed from the melt. If care is taken in the preparation of the polymer sheet, ie, if nucleation takes place in contact with a high-energy substrate to prevent any mechanical work upon removal of the polymer from the substrate, then it is possible to prepare this polymer for adhesive bonding without resorting to crosslinking of the surface region.

Surface Morphology and Wettability

Although considerable effort has been expended in determining the relationship between the contact angle of a liquid on a polymer surface and the chemical constitution of that surface, little attention has been paid to the detailed physical properties (eg, molecular-weight distribution, density, crystallization behavior, etc) of the polymer and, more important, the detailed procedure for preparing the polymer surface for wettability studies. In general, there is rather widespread agreement in the accepted values for the critical surface tensions of wetting, γ_c, for a variety of polymeric species. In most tabulations, the wettability of a species is related to the presence of a particular functional group that resides in the outermost surface layer (Table 2).

In this section a more critical approach to the concept of wettability will be taken. The initial remarks are confined to polyethylene but will be extended to a variety of polymers. See also Effect of Chemical Constitution.

Table 2. Critical Surface Tension of Wetting (γ_c) of Polymers at 20°C

Polymer	γ_c, dyn cm^{-1}
polyethylene	31
polystyrene	33–35
polytetrafluoroethylene	18.5
polyhexafluoropropylene	16.2–17.1
polychlorotrifluoroethylene	31
poly(vinyl alcohol)	37
poly(methyl methacrylate)	33–44
poly(hexamethylene adipamide)	46
polydimethylsiloxane	24

Surface Density. Although polyethylenes may have a density range of 0.855 g-cm^{-3} for the amorphous polymer to 1.00 g-cm^{-3} for the completely crystalline species (see Polyethylene under ETHYLENE POLYMER), until recently no significant differences have been reported in their γ_c values. The density range noted above is comparable to the difference between hexane and hexadecane, two hydrocarbons which differ considerably in their surface chemical properties. Apparently, there is some feature common to all polyethylenes which tends to equate their wettabilities.

Comparisons of extrapolated values of the melt surface tension of several nonpolar polymers (Table 3) and their commonly accepted γ_c values (Table 2) are striking. These data led the author to postulate that $(\gamma_{LV})_p^{20} = \gamma_c$ for melt-crystallized nonpolar polymers, where $(\gamma)_p^{20}$ is the surface tension of the molten polymer extrapolated back to room temperature. Obviously, when $(\gamma_{LV})_p^d \neq (\gamma_{LV})_p$ where $(\gamma_{LV})_p^d$ is the dispersion force contribution to the surface tension, the above relationship is no longer valid. These dispersion forces are the results of the interaction of fluctuating electric dipoles with the induced dipoles; they contribute to cohesion in all substances, but their magnitude depends on the type of material and its density. From this comparison the surface layer of the melt crystallized polymers listed in Table 3 is concluded

Table 3. Extrapolated Melt Surface Tensions of Polymers

Polymer	$(\gamma_{LV})_p^{20}$, dyn cm^{-1}
polyethylene	36.2
polypropylene	28.0
polychlorotrifluoroethylene	30.8
polydimethylsiloxane	20.6
polystyrene	32.4

to be essentially amorphous. Investigators have shown that to obtain reasonable agreement between γ_c and the parachor, the amorphous density of the polymer has to be used. Apparently, all polyethylenes, when prepared in the conventional manner, behave with respect to wettability as if they have a surface layer which is essentially that of the supercooled liquid. Although their bulk densities may vary over a considerable range, their surface densities are similar. This would account for their similar wettability behavior. Recently, it has been shown that a well-characterized preparation of polyethylene single crystal aggregates has a γ_c of approximately 54 dyn/cm, considerably in excess of the commonly accepted value (Table 2). In this instance, the surface density of the polymer is no longer 0.855 g-cm^{-3} but is computed to be approximately 0.94 g-cm^{-3}. The low value of 0.94 g-cm^{-3} for the surface density of single crystal aggregate may result from the defects presumed to be present in the fold structure.

It will be shown in later sections, employing a wettability approach, that the morphological structure of the surface region of melt-crystallized polymers, particularly polyethylene, is strongly dependent upon the manner in which this region is generated.

Effect of Temperature. While a considerable effort has been devoted to measuring the contact angle of a variety of liquids on surfaces at ambient temperatures, only a minor effort has been directed to the temperature dependence of wetting. It has been observed that the contact angle for a variety of liquids on polymers is essentially invariant with temperature, although it is to be expected that there would be a decrease at quite elevated temperatures.

The several attempts to analyze the wettability data at elevated temperatures have been made with the Fowkes-Young equation (eq. 3). If it is assumed that $\gamma_{SV}^d = (\gamma_{LV})_p^d$, where p refers to polymer, for melt-crystallized polymers that are nucleated and solidified in contact with a low-surface-energy phase, and that the ratio $\gamma_{LV}^d/\gamma_{LV}$ for the wetting liquid is constant, then,

$$\cos \theta = 2 \left[\frac{k(\gamma_{LV})_p^d}{\gamma_{LV}} \right]^{1/2} - 1 \qquad (3)$$

where $k = (\gamma_{LV}^d/\gamma_{LV})$ and θ is the contact angle.

A knowledge of the variation of the surface tension of both the wetting liquid and the polymer as a function of temperature is sufficient to permit a calculation of θ as a function of temperature. Since $(\gamma_{LV})_p^d/\gamma_{LV}$ is relatively insensitive to the increase in temperature, the value of θ is essentially constant. However, this is probably not general with respect to all liquids in contact with low-surface-energy polymers. Since the test liquid and polymer probably have different critical temperatures, ie, the temperature at which $\gamma_{LV} = 0$, there should be an intermediate range, assuming no degradation of the polymer, where $(\gamma_{LV})_p = \gamma_{LV}$. When this condition is fulfilled, $\theta = 0$. For example, the critical temperature of polyethylene has been estimated to be $1031°K$, which is considerably higher than any of the liquids normally used in wettability studies.

Interfacial Contact. Generally, to facilitate ease of removal from molds, polymers have been prepared for wettability studies by pressing or solidifying them in contact with low-surface-energy solids (eg, polytetrafluoroethylene) or with high-energy surfaces for short times at low temperatures (ie, just above the melting point of the polymer). Experience has shown that if longer periods of time are used at relatively high temperatures on high-energy surfaces, it is difficult to remove the polymer from the metal surface without damaging the surface layer of the polymer. Longer dwell times of the polymer melt on the surface of the metal results in the formation of a strong adhesive joint, provided that the metal does not have a weakly adherent oxide layer. If a mechanically weak oxide layer were present, it is conceivable that this could be transferred to the polymer, thereby affecting the wettability results.

Recent studies of the kinetics of wetting of polymer melts on surfaces have shown that the ability of a polymer melt to attain an equilibrium contact angle with a solid substrate is proportional to the surface tension of the polymer melt (γ_{LV}) and inversely proportional to the melt viscosity (η). Since the viscosity varies greatly with temperature, to preclude interfacial voids in the polymer melt–high-energy surface (ie, metal, metal oxide) interface, at low temperatures it is important to allow sufficient time to insure complete mating of the surfaces (Fig. 2a indicates poor wetting). Although the thermodynamic requirements for spreading are fulfilled (ie, $\gamma_{SV} \geq \gamma_{LV} + \gamma_{SL}$), spreading of the polymer may not take place because of the kinetic requirements. Enhancement of wetting is rather easy to accomplish by employing higher temperatures and longer times. Since η decreases markedly with increasing temperature, relatively short times are required at higher temperatures to achieve extensive mating of the liquid and high-energy solid (Fig. 2b).

Effect of Substrates on Surface Region Morphology of Polyethylene. Surface studies on crystallizable polymers, eg, polyethylene, have ignored, in general, the nature of the nucleating phase (ie, vapor, solid, or liquid) and the details of formation of the polymer *melt*–nucleating phase interface which, on soldification by cooling, results in a polymer *solid*–nucleating phase interface.

Extensive heterogeneous nucleation of polyethylene melts on high-energy surfaces results in generation of transcrystallinity in the interfacial region $[(S–L) \rightarrow (S–S)]$ (Fig. 4). Investigators have observed that there is a variation in the extent of supercooling which may depend upon surface energy and interatomic spacing in the substrate. Effective nucleating agents allow for only small supercoolings. Others have concluded that stresses at the interface set up during cooling from the melt are important in determining the subsequent morphology.

Low-energy surfaces (eg, polytetrafluoroethylene) are apparently ineffective nucleating agents. When polymers are cooled in contact with these surfaces, nucleation is precluded at the solid–liquid interface and is apparently initiated in the bulk. Sufficient supercooling has not occurred at the solid–liquid interface to nucleate the interfacial region before nucleation occurs in the bulk. Apparently, this is the reason for the lack of a well-defined transcrystalline region when polyethylene is nucleated against a low-energy solid. As crystallization proceeds in the bulk, polymer molecules that cannot be accommodated into the crystal lattice during crystallization are rejected to the interface.

Employing high-energy surfaces for the nucleation of a polymer melt is effective only if sufficient time is allowed for the polymer melt to achieve extensive and intimate contact with the substrate. This, as mentioned earlier, is a kinetic requirement. If sufficient time has not been allowed, considerable interfacial voids will result and nucleation will generally occur in the bulk. If sufficient time is allowed for spreading to occur, a situation similar to that of Figure 2b will result, where interfacial voids are precluded and nucleation occurs predominantly at the *S–L* interface. The mere presence of a high-energy surface does not, in itself, ensure that intensive and intimate contact will occur, and that a highly nucleated surface region will result upon solidification of the polymer melt.

Wettability and Surface Morphology of Polyethylene. Since the morphology of polyethylene is strongly dependent upon the polymer melt–substrate system, the effect of the substrate on the wettability of polymers was studied. To avoid damaging the surface region of the polyethylene (after solidification) by mechanical means, dissolution techniques were used to separate the polyethylene from the substrate. For all the substrates listed in Table 4, except polytetrafluoroethylene, gold, and the salt crystals, the substrates were dissolved in either concentrated hydrochloric acid or sodium hydroxide at temperatures below 30°C. The gold was dissolved by either amalgamation with mercury or exposure to a concentrated aqueous sodium cyanide solution. There is enough dissolved oxygen in the aqueous phase to oxidize the gold and enable it to be complexed. The salt crystals were dissolved by exposing the solid polyethylene–crystal system to water. The polyethylene–polytetrafluoroethylene composite separated by itself on cooling from the melt.

The polyethylene specimens were prepared by placing a piece of polymer ($\sim$10 mil thick) onto an evaporated metal film (deposited on glass microscope slide, 3 in. $\times$ 1 in. $\times$ 1/16 in.), and placing them in an oven at 200°C in a nitrogen atmosphere. The composites remained in the oven for a period of 30 min, an interval sufficient to allow for extensive contact between the polymer melt and the substrate. One surface of the polymer melt was in contact with the vapor (nitrogen) to provide a control. In all cases, the vapor-nucleated polyethylene surface gave contact angles with glycerol ($\theta = 81°$) which were similar to those reported by Zisman (4). Dissolution of the substrates from the solidified polymer did not alter the contact angle of glycerol on the vapor-nucleated surface. After dissolution, no residue of the substrates was detected by conventional analytical techniques (atomic absorption spectroscopy, electron microprobe, etc).

The wettability results in Table 4 are striking. For polyethylene nucleated and solidified in contact with nitrogen or polytetrafluoroethylene, the glycerol contact angle is similar to that reported by Zisman (4). For polyethylene nucleated and solidified on a variety of high-energy surfaces, except copper, a marked decrease in θ is noted. Films generated against these high-energy surfaces were examined, using alternated total reflectance infrared techniques, to determine if oxidation was taking place. No evidence for oxidation was found. Gold, mercury, and the salt crystals produced polyethylene surfaces with the lowest glycerol contact angle. If oxidation were taking

Table 4. Wettability of Melt-Crystallized Polyethylene Film Formed by Nucleation at Solid–Liquid and Liquid–Vapor Interfaces[a]

Surface	θ, deg	$\cos \theta$	$\rho_S^{a,ac,c}$ g/cm³	$V_{sp}^{a,ac,c}$ cm³/g	$\gamma_{LV}(\gamma_c)$, dyn/cm	% crystallinity
nitrogen	81	0.1564	0.855	1.1695	36.2	0
polytetrafluoroethylene	81	0.1564	0.855	1.1695	36.2	0
Mylar polyester	81	0.1564	0.855	1.1695	36.2	0
copper	80	0.1737	0.862	1.1601	37.4	5.1
nickel	68	0.3746	0.933	1.0718	51.3	53.3
tin	66	0.4067	0.944	1.0593	53.8	60.1
aluminum	65	0.4226	0.949	1.0537	54.9	63.2
glass	65	0.4226	0.949	1.0537	54.9	63.2
chromium	64	0.4384	0.954	1.0482	56.1	66.2
mercury	57	0.5446	0.989	1.0111	64.8	86.4
gold	53	0.6018	1.007	0.9930	69.6	93.6
tantalum	60	0.5000	0.975	1.0256	61.4	78.5
NaCl	57	0.5446	0.989	1.0111	64.8	86.4
KBr	57	0.5446	0.989	1.0111	64.8	86.4
KCl	57	0.5446	0.989	1.0111	64.8	86.4
CaF₂	55	0.5736	0.998	1.0020	67.3	91.4

[a] Polyethylene was in contact with surfaces for 30 min at 200°C. All measurements were performed at 20°C.

[b] a, amorphous; ac, partially crystalline; c, crystalline.

place it is reasonable to assume it would occur more readily when the polymer melt was in contact with a metal oxide. A more conclusive argument will be presented later, when the wettability behavior of a variety of thermoplastic polymers nucleated and soldified against gold will be described.

Apparently, as a result of extensive nucleation of molten polyethylene on high-energy surfaces, there is generated a plane of spherulitic cores which exhibit quite low contact angles with glycerol (Fig. 8). When polyethylene is nucleated against a low-surface-energy phase (polytetrafluoroethylene or nitrogen) or a high-surface-energy phase that is poorly wetted with the polymer melt, thus forming a low-energy, gas–melt interface, all polyethylenes examined exhibit the same contact angle with glycerol regardless of their bulk densities.

Recently, the Fowkes-Young equation has been modified to account for possible differences in the surface density of polymers. Using a similar analysis for polyethylene nucleated on a variety of substrates, the values of the surface density were computed. It can be shown that

$$(\cos \theta)_{a,ac,c} = \frac{2\left(\dfrac{\rho_S^{a,ac,c}}{\rho_S^a}\right)^2 [(\gamma_{LV}^d)_p \gamma_{LV}^d]^{1/2}}{\gamma_{LV}} - 1 - \frac{\pi_e}{\gamma_{LV}} \qquad (4)$$

where θ is the contact angle of the sessile drop of liquid on the polymer surface; ρ_S is the surface density; γ_{LV} is the surface tension; the subscript p refers to the polymer; the superscript d refers to the dispersion component of the surface free energy; the superscripts a, ac, and c refer to amorphous, partially crystalline, and crystalline, respectively; and π_e is the spreading pressure $(\gamma_S - \gamma_{SV})$, ie, the difference between the surface free energy of the surface of the polymer in vacuum and in equilibrium with the vapor of the wetting liquid. We have ignored π_e/γ_{LV} in the computations since $\theta > 0$.

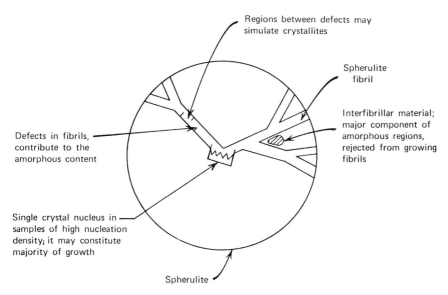

Regions between defects may simulate crystallites

Spherulite fibril

Interfibrillar material; major component of amorphous regions, rejected from growing fibrils

Defects in fibrils, contribute to the amorphous content

Single crystal nucleus in samples of high nucleation density; it may constitute majority of growth

Spherulite

Fig. 8. Model of spherulite.

When the surface density of the solid is essentially that of the amorphous solid or supercooled liquid, equation 4 becomes the familiar Fowkes-Young expression. Based on the generally accepted value of ρ_S^a for polyethylene, ρ_S^{ac} values are computed. From these values, the specific volumes are obtained. The percent crystallinity in the surface region is computed from the specific volume where

$$\% \text{ crystallinity} = 100 \left(\frac{V_{sp}^{ac} - V_{sp}^{a}}{V_{sp}^{c} - V_{sp}^{a}} \right) \tag{5}$$

The variation of $\cos \theta$, θ, and γ_c as a function of polymer melt–substrate (aluminum, gold, polytetrafluoroethylene, and nitrogen) contact time is shown in Figure 9. It is of interest to note that the γ_c of the polymer generated against both polytetrafluoroethylene and nitrogen is essentially constant. The values of $\cos \theta$ for both aluminum and gold approach the limiting values recorded in Table 4. For shorter times, a spectrum of γ_c values can be computed. This probably reflects the poor extent of wetting for shorter times, resulting in a situation such as Figure 2a, where nucleation occurs both at the metal and in the bulk phase. A surface layer is formed which is less dense than one where extensive contact has been achieved (Fig. 2b).

A comparison among the contact angles of a variety of liquids on polyethylene surfaces generated against polytetrafluoroethylene and gold is presented in Table 5. Included with these data are the wettability results obtained for the polyethylene single crystal aggregates. The gold-nucleated polyethylene is considerably more wettable than the polyethylene single crystal aggregates. In Figure 10, a Fowkes type plot of the data in Table 5 is constructed. The value of γ_S^d for the gold-nucleated surface is 69.6 dyn/cm, considerably in excess of the values 53.6 dyn/cm and 36.2 dyn/cm for the polyethylene single crystal aggregate and the polytetrafluoroethylene-nucleated polymer, respectively. It appears that γ_c varies with the substrate used in preparing the polyethylene surface. Apparently for polyethylene we may have 36.2 dyn/cm $\leq \gamma_c \leq$ 69.6 dyn/cm.

Table 5. Wettability of Polyethylene at 20°C

Liquid	γ_{LV}, dyn/ cm	γ_{LV}^{d}, dyn/ cm	$(\gamma_{LV}^{d}/\gamma_{LV})^{1/2}$, (dyn/cm)	Single crystal		Nucleated against			
						polytetrafluoro- ethylene		gold	
				θ, deg	$\cos\theta$	θ, deg	$\cos\theta$	θ, deg	$\cos\theta$
water	72.8	21.8	0.0641	93	−0.052	94	−0.070	84	0.105
glycerol	63.4	37.0	0.0959	67	0.391	79	0.191	53	0.602
formamide	58.2	39.5	0.1080	55	0.574	77	0.225	41	0.755
α-bromo- naphthalene	44.6	44.6	0.1497	spreads	1.000	35	0.818	spreads	1.000

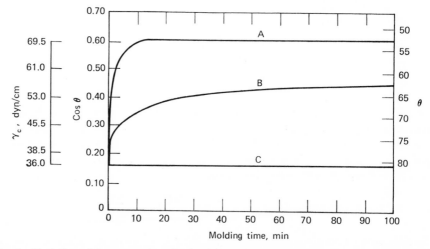

Fig. 9. Variation of θ, $\cos\theta$, and γ_c obtained from the wetting of glycerol on polyethylene which was molded for varying lengths of time at 200°C on a variety of surfaces in a nitrogen atmosphere. Metal substrates removed by dissolution. Polytetrafluoroethylene composite separates on cooling. A, gold; B, aluminum (etched); C, nitrogen and polytetrafluoroethylene.

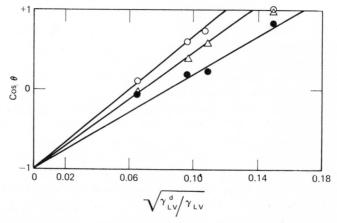

Fig. 10. Typical Fowkes plot of wettability data for several polyethylenes. ●, nucleated in N_2 or molded against low-energy surface; △, single crystal aggregate; ○, gold nucleated.

Surface Density of Polymers Nucleated on Gold. One argument against possible contamination of the polymer surface when in contact with gold is that atactic polypropylene (amorphous) exhibits the same contact angle with glycerol whether it has been formed at the polymer melt–gold interface or at the polymer melt–vapor interface (Table 6). In addition, poly(4-methyl-1-pentene), whose crystal density is reported to be less than the amorphous density, shows a slight increase in the contact angle for the surface generated at the polymer melt–gold interface. If surface contamination were substantial, increases in wettability would also be observed with these polymers.

Based on the extrapolation of the melt surface tension data for polyethylene, polypropylene, polychlorotrifluoroethylene (Table 2) and the suggestion that $\gamma_{LV} = \gamma_c$, values of $(\gamma_{LV})_p$ for polymers in Table 6 were estimated. For all the polymers in Table 6 but nylon-6, $\gamma_{LV}^d = (\gamma_{LV})_p = \gamma_c$. For nylon-6, the value of $(\gamma_{LV})_p$ was computed for a $60°$ glycerol contact angle which is representative of the amorphous surface. Employing equation 4 and the values for $(\gamma_{LV})_p$, θ, ρ_S^a, γ_{LV}^d of glycerol, and neglecting the contribution of π_e/γ as suggested by Fowkes, ρ_S^{ac} values were computed.

The somewhat high value of ρ_S^{ac} for polychlorotrifluoroethylene may be either a result of employing a fourth power dependence of the density in deriving equation 4 or a low estimate for the literature value of ρ_S^c.

In measuring γ_c, investigators have agreed generally to certain values which are representative of the constitution of a polymeric species (Table 2). This has been related to the presence of a particular functional group. The data in Table 6 indicate that the wettability of a crystallizable polymer can be made to vary within limits by controlling the nature of the heterogeneous nucleation and subsequent formation of the surface region of the polymer. Heterogeneous nucleation of certain polymers on high-energy surfaces can lead to surface densities, as computed from wettabilities, which are similar to the single crystal densities.

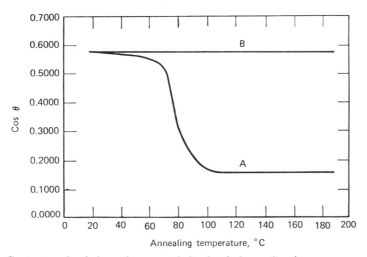

Fig. 11. Contact angle of glycerol on annealed polyethylene. Specimens were annealed a minimum of 5 hr at the cited temperatures in a nitrogen atmosphere. A, Films nucleated against gold, cooled to 20°C then gold dissolved. Films annealed at specified temperatures; B, Films nucleated against gold, cooled to 20°C. Films in contact with gold were annealed, lowered to 20°C then gold removed by dissolution.

Table 6. **Wettability Data for Polymer–Glycerol Systems at 20°C**

Polymer	Bulk density, g/cm³	Temperature of preparation, °C	θ_{MV},[a] deg	θ_{Lit}, deg	θ_{Au},[b] deg	γ_{LV}, dyn/cm	γ_c, dyn/cm	ρ_s^a, g/cm³	V_{sp}^a, cm³/g	ρ_s^c, g/cm³	V_{sp}^c, cm³/g	$\rho_s^{ac,c}$, g/cm³	V_{sp}^{ac}, cm³/g	$(\gamma_c)Au$, dyn/cm	% crystallinity in surface region
polyethylene	0.95	190	81	81	53	36.2	35	0.855	1.1695	1.014	0.9862	1.007	0.993	69.6	96.3
nylon-6	1.14	280	60	60	35	46	46	1.09	0.9174	1.24	0.8065	1.23	0.813	74.4	94.1
polychlorotrifluoroethylene	2.12	230	86	86	62	30.8	31	1.925	0.5195	2.19	0.4566	2.264	0.4417	58.9	100
polypropylene (isotactic)	0.90	220	92	92	78	28.0	29	0.855	1.1695	0.930	1.0753	0.932	1.0729	39.5	100
polypropylene (atactic)	0.86	220	92	92	92	28.0	29	0.855	1.1695	0.930	1.0753	0.855	1.0729	28.0	0.0
poly(4-methyl-1-pentene)	0.83	230	95		96	22.6		0.838	1.1933	0.827	1.2092	0.827	1.2092	22.0	100

[a] Contact angle of glycerol on a surface generated at the polymer melt–vapor interface.
[b] Contact angle of glycerol on a surface generated at the polymer melt–gold interface.
[c] Computed from equation 4.

Stability of Crystalline Surface Region. Stability of the surface with respect to temperature was examined to determine if the surface was subject to reorganization. Composites consisting of gold–polyethylene–gold were formed at 200°C for ½ hr in a nitrogen environment. The effect of annealing at elevated temperatures on the film before and after dissolution of the gold was followed by measuring the contact angle of glycerol on these surfaces. Figure 11 shows these data. If the film is annealed in contact with the gold, the observed contact angle is unchanged. If the composite is stripped of gold (by dissolution, as before) and then the films are annealed, there is a marked change in θ. The contact angle approaches that of the polytetrafluoroethylene-nucleated film. Apparently, there is some surface melting of the small crystallites formed at the solidified polymer melt–gold interface even at these low temperatures. If the polymer is in contact with the gold during the annealing, the adsorbed species apparently have insufficient mobility to reorganize.

Bibliography

1. H. Schonhorn and L. H. Sharpe, *J. Polymer Sci.* [B] **2,** 719 (1964).
2. L. H. Sharpe and H. Schonhorn, in *Contact Angle, Wettability and Adhesion,* No. 43 in *Advances in Chemistry Series,* American Chemical Society, Washington, D. C., 1964, p. 18.
3. H. Schonhorn and L. H. Sharpe, *J. Polymer Sci.* [A] **3,** 3087 (1965).
4. W. A. Zisman, in Ref, 2, p. 1.

General

R. Houwink and G. Salomon, eds., *Adhesion and Adhesives,* 2nd ed., Elsevier Publishing Co., New York, 1965.
D. D. Eley, ed., *Adhesion,* Oxford University Press, 1961.
J. J. Bikerman, *The Science of Adhesive Joints,* 2nd ed., Academic Press, New York, 1968.
I. Skeist, ed., *Handbook of Adhesives,* Reinhold, New York, 1962.
P. Weiss, ed., *Adhesion and Cohesion,* Elsevier, Amsterdam, 1962.
F. P. Bowden and D. Tabor, *The Friction and Lubrication of Solids,* Part I, 1st ed., 1950; Part II, 1st ed., 1964; Clarendon Press, Oxford University Press.
Contact Angle, Wettability and Adhesion, No. 43, in *Advances in Chemistry Series,* American Chemical Society, Washington, D.C., 1964.
R. L. Patrick, ed., *Treatise on Adhesives and Adhesion,* Marcel Dekker, New York, 1966.
A. Sharples, *Polymer Crystallization,* St. Martin's Press, New York, 1966.

Harold Schonhorn
Bell Telephone Laboratories, Inc.

ADHESIVE COMPOSITIONS

This section deals with adhesives as chemical compositions. An *adhesive* is defined by the American Society for Testing and Materials (ASTM) as a substance capable of holding materials together by surface attachment (1). The term *adherend* is generally used to refer to the body held to another body by the adhesive.

Adhesives offer many advantages over other methods of joining materials (2–10): (*1*) Thin films and small particles that could not be joined by other techniques are readily bonded with adhesives. Typical of such products are labels, abrasive wheels, sandpaper and emery cloth, laminates of plastic film and aluminum foil, nonwoven fabrics, particle board, and veneered furniture. (*2*) Stresses are distributed over wider areas, making possible assemblies lighter than could be achieved by mechanical fastening. Jet aircraft may be constructed of sandwich panels in which thin faces of aluminum are bonded to a honeycomb core. Aerodynamic drag is reduced because there are no protruding rivets. (*3*) Dimensional stability of anisotropic materials can be improved by crossbonding, as in plywood and random-web nonwoven fabrics. (*4*) The layer of glue acts also as an insulator in electrical equipment, a moisture barrier in laminates and sealants, and a corrosion-inhibiting barrier when two dissimilar metals are bonded together. (*5*) Most important of all, in many applications, is the lower processing cost. Weaving of cloth, sewing of leather, soldering, brazing or welding of metals, and mechanical fastening with rivets, bolts, or nails are some of the joining methods that can be economically replaced by adhesive-bonding methods.

Economics

Adhesives are essential components needed to make hundreds of products, among them aircraft, automobiles, corrugated cartons, plywood, envelopes, stamps, pressure-sensitive and remoistenable tapes, abrasive wheels, laminates, labels, shoes, tires, and nonwoven fabrics.

Consumption of adhesives in the United States has been growing at the annual rate of 7%, considerably faster than the gross national product (11). Table 1 lists the 1960 prices and pound and dollar value for various adhesives; Table 2 shows the breakdown of adhesives usage in industry. The figures are on a dry polymer basis, and are all at the raw-material supplier's level, except for those on the rubber-based adhesives, which are at the formulator's level. Starch and dextrin are seen to be the largest-volume adhesives, but the rubber-based adhesives have the greatest dollar value. An analysis of adhesives usage over a five-year period reveals that the rubber cements are increasing most rapidly in terms of dollar value, especially in pressure-sensitive tapes, automobile bodies, shoes, and building applications. On a percentage basis, the fastest growing materials are epoxy resins, for structural bonding; poly-

61

Table 1. Total Adhesives Consumption in 1960 (11)

Adhesive	Consumption, millions of pounds	Price, dollars/pound	Consumption, millions of dollars	% of total value	% of total volume
starch and dextrin	575	0.07	40.3	9.1	26.8
soya	84	0.10	8.4	1.9	3.9
animal glue	98	0.20	19.6	4.4	4.6
blood	55	0.12	6.6	1.5	2.6
casein	10	0.20	2.0	0.5	0.5
amino resin	156	0.15	23.4	5.3	7.3
phenolic resin	254	0.28	71.1	16.0	11.9
epoxy resin	5	0.65	3.3	0.7	0.2
rubber	210	0.90	189.0	42.5	9.8
polysulfide polymer	10	1.00	10.0	2.3	0.5
poly(vinyl acetate)	80	0.35	28.0	6.3	3.7
natural resin and derivatives	10	0.25	2.5	0.6	0.5
petroleum resin	35	0.15	5.3	1.2	1.6
cellulosic plastics	4	0.45	1.8	0.4	0.2
asphalt	400	0.025	10.0	2.3	10.6
silicates	115	0.033	3.8	0.9	5.4
others	40	0.45	18.0	4.1	1.9
total, approx	2150		450	100.0	100.0

sulfide rubbers, for use as sealants in curtain-wall construction; and poly(vinyl acetate), for packaging and household adhesives.

Application and Setting

In order to have good bonding between adhesive and substrate, there must be a decrease in free energy when the two are brought into contact. In addition, the adhesive must be sufficiently fluid to wet the surface completely. To meet this requirement, the adhesive must be low in viscosity at the time of application. Eventually, however, adhesives must become sufficiently high in viscosity, even "infinitely" viscous, through crosslinking or crystallization, so that their cohesive strength, ie, the strength binding the adhesive together, will be high.

The *setting* of an adhesive (its transformation from a fluid to a solid) may take place in several ways.

Cooling of a Hot Melt. The hot melts are at once the oldest and newest of adhesives. Bitumen (see BITUMINOUS MATERIALS) was supposedly the mortar for the Tower of Babel; beeswax and pine tar were used in calking the Roman vessels that dominated the Mediterranean Sea. Asphalt, as cheap as a penny a pound, is favored today as a waterproofing laminate for multiwall paper bags and military packaging, as well as a binder of aggregates in roads. Paraffin wax, now often fortified with microcrystalline wax and/or polyethylene, is an economical and effective coating for breadwraps and other food packaging; it retards the transmission of moisture vapor and permits heat seals to be made in high-speed packaging equipment. Greater bond strength is obtained from cellophane by coating it with a thin film of plasticized cellulose nitrate or saran. Vinyl acetate polymers compounded with solid adhesives comprise the coating for "delayed tack" labels, which remain sticky for many seconds after exposure to heat, thus facilitating high-speed application to bottles and cans.

Table 2. Consumption of Adhesives by Industry in 1960 (11)

Use	Million lb	Million $
corrugated	315	25
case sealing	125	10
paper bags and envelopes	71	5.5
tapes and labels	63	5
folding and set-up boxes	110	9
fiber drums and tubes	70	5.5
pressure-sensitive tapes	50	40
bookbinding	60	12
plywood	270	50
particle board	44	13
hardboard	14	3
lumber and timber	10	2.5
wood furniture	25	6
miscellaneous wood products	10	2
automotive	67	30
tire cord	23	7
bonded abrasives	15	4
other metal	5	4
shoes and other leather products	40	30
building	570	78
household	32	57
nonwoven fabrics	32	6.5
others	140	42.0
total, approx	2150	450

Milk cartons, long a marketing domain for wax-coated papers, are being made increasingly from polyethylene-coated paperboard. Polyethylene itself has vast usage as a heat-sealable packaging film. It therefore acts as its own hot-melt adhesive. Polyethylene film also can be bonded with formulations in which polyethylene, polyisobutylene, and low-molecular-weight hydrocarbon resins are blended. Copolymers of ethylene with 18–28% vinyl acetate are also rapidly achieving acceptance in this field. When they are blended with 2–4 times their weight of paraffin wax, they are conveniently and effectively adapted to automated packaging operations.

Safety glass is made by sandwiching a highly plasticized film of poly(vinyl butyral) between two sheets of plate glass with heat and pressure. Ethylcellulose and cellulose acetate butyrate are among the thermoplastic resins that have been formulated with plasticizers and waxes to provide low-melting-point dip baths for the sealing of packages against moisture penetration. Patches of cloth that have been coated with vinyl plastisol can be applied with a hot iron to repair tears in clothing and sheets.

Among the strongest bond formers of the newer hot-melt adhesives are polyamides made by condensation of diamines and "dimer acids" derived from unsaturated fatty acids (see ACIDS AND DERIVATIVES, ALIPHATIC). The amide groups on these compounds provide good adhesion to paper, leather, metals, and rubber, and the intervening aliphatic groups impart toughness to the adhesive joint.

Hot-melt adhesives can be made fluid by heating in a large feed-stock vessel, the reservoir. They are then applied to the substrate with the aid of rolls or pumps. Two new devices have become available that avoid the long exposure of the adhesive to elevated temperatures of the reservoir technique. Both heat the adhesive only as fast as it is to be used. In one, granules of adhesive are fed into a heater–extruder with a short barrel. In the other device, the adhesive is made available as a coiled rope; its end is fed to a heated melting wheel that applies it, liquefied, to the substrate surface.

Because of their uses and methods of application, hot melts must have suitable *tack* characteristics, more narrowly controlled than other types of adhesives. Tack is defined by ASTM as "the property of an adhesive that enables it to form a bond of measurable strength immediately after adhesive and adherend are brought into contact under low pressure" (2). To have good tack, the newly formed joint must possess both adhesion and cohesion in at least moderate values. The adhesive used must be fluid enough to wet the substrate, yet viscous enough to resist separation within the film it forms (the "glue line"). In practice, it has been found that best tack characteristics usually occur in the viscosity range of 10^4–10^6 cP.

Waxes generally have too low a viscosity in their molten state for good cohesive strength, whereas unmodified thermoplastic polymers of high molecular weight have to be heated to too high a temperature in order to make them adequately fluid.

Release of Solvent. Water and organic solvents are convenient carriers and viscosity reducers for adhesive compositions. Water is the solvent for starches and dextrins, protein glues, sodium silicates, poly(vinyl alcohol), and several of the thermosetting resins. In these materials, viscosity is directly related to molecular weight; hence the solids content used can seldom reach 50% if excessive viscosities are to be avoided. On the other hand, in rubber latexes and poly(vinyl acetate) emulsions, the polymer is in the dispersed phase, and the viscosity is controlled chiefly by varying the composition of the continuous phase. Consequently, it is possible to have high fluidity despite solids concentration of 35–55% and molecular weights in the hundreds of thousands, or higher.

An important adhesive for corrugated boxboard consists of starch granules suspended in an alkaline solution of dextrin, with a solids content higher than 60%. High solids content of the adhesive is a vital consideration, since removal of water is the most time-consuming step of the process.

Organic solvents are used in *cements* formed from natural or reclaimed rubber, neoprene, nitrile rubber, cellulose nitrate, cellulose esters, or some vinyls. Solids concentration is usually 12–35%. Molecular weight is characteristically low. "Half-second" cellulose nitrate (the time refers to flow time in the standard viscosity test) and cellulose acetate butyrate are interesting examples of short-chain polymers that are often found satisfactory as adhesives but are too weak for most, if not all, plastic products.

In the bonding of impervious substrates, the solvent-containing cement is applied to the fay surfaces and the solvent is allowed to evaporate before completing the joint. *Contact cements*, for example, are rubbery compositions sufficiently tacky after evaporation of solvent to form a strong bond with only contact pressure.

One of the most important groups of solvent-based adhesives are the *pressure-sensitive* coats, which are applied to paper, cloth, and other web materials in the manufacture of tapes and labels. A pressure-sensitive adhesive is one that will adhere itself to a surface upon application at room temperature with only lightly applied pressure. Pressure-sensitive tapes are useful for masking, packaging, insulating, sealing, labeling, bundling, and mending, as well as for a variety of household and medical purposes.

Natural, reclaimed, and styrene–butadiene rubbers are the predominant polymers in pressure-sensitive adhesive formulations. For special applications, where higher costs are acceptable, polyisobutylenes, poly(vinyl ethers), and poly(acrylic esters) are also finding uses. In addition to the elastomer resin and solvent, pressure-sensitive formulations often include: resin tackifiers, such as rosin esters, oil-soluble phenolics, or polyterpenes; antioxidants; plasticizers, such as mineral oil, liquid polyiso-butylenes, or lanolin; and fillers, such as zinc oxide or hydrated alumina.

Aliphatic and aromatic hydrocarbons are the preferred solvents for natural and reclaimed rubber cements. Addition of ketones may be necessary to dissolve the more polar synthetic rubbers. The cellulosic polymers and some of the vinyl resins require the use of ketones and esters.

Cost, toxicity, and flammability are among the formulator's and user's objections to organic solvents in adhesives. If the user's operation is large enough to justify it, solvents can be recovered. Benzene, once common as a rubber-cement solvent, is now in disfavor because of its cumulative toxicity. Flammability can be minimized or eliminated through the use of chlorinated solvents but at an increase in cost.

Organosols are dispersions of poly(vinyl chloride) resins in plasticizers and volatile organic liquids. After the organosol-type adhesive has been applied to paper, cloth, or metal, the coating is heated to rapidly evaporate the solvent and fuse the resin to effect a bond.

Polymerization in Situ. This category of adhesive application and setting includes all of the thermosetting resins. Many of these, eg, urea–formaldehyde melamine–formaldehyde (see AMINO RESINS), phenol– and resorcinol–formaldehyde resins (see PHENOLIC RESINS), are applied with water or sometimes alcohol as the solvent or dispersing medium. Nitrile and neoprene rubbers, applied from organic solvent or aqueous latex, can belong in this classification if they are compounded with vulcanizing agents to enable them to develop higher cohesive strength when heated. Other polymerizing-in-situ adhesives include the epoxy resins (qv), the isocyanate polymers, and vinyl monomers, such as methyl methacrylate and methyl 2-cyano-acrylate (see ACRYLIC ESTER POLYMERS).

In all these materials, a chemical reaction takes place in the layer of adhesive, the "glue line," after the two adherends have been brought together. The polymerization results in building up the size of the polymer molecule and thus developing bond strength. The adhesives may be divided into two groups. The first group are those that form high polymers by condensation, usually with water as a by-product. This group includes the phenolic and amino resins. They are important for plywood and other wood-adhesion uses. The second group are the adhesives that cure or poly-merize without formation of by-products. Included in this category are the epoxy resins, isocyanate polymers, acrylic monomers, and rubbers.

In many instances, the adhesives can be cured without external application of heat. Examples are: epoxy resins, isocyanate resins, some acrylic monomers, un-saturated polyesters, and resorcinol–formaldehyde resins. The polymerization re-action is usually exothermic, so that the temperature of the glue becomes higher than the ambient temperature of the substrate; this is especially the case with adherends, such as wood, that conduct heat poorly.

As a group, the thermosetting adhesives yield products that have superior strength, dimensional stability, and resistance to elevated temperature, to creep, to organic solvents, and to water, as compared with hotmelts, pressure-sensitive adhesives, and rubber cements.

Adhesive Materials

The organic adhesives are all high polymers. The oldest are those of natural origin. Animal glues and starches have been in use since ancient times; cements based on natural rubber and pyroxylin (cellulose nitrate) became available during the 1800s. Resins based on phenol–formaldehyde and urea–formaldehyde were developments in the early twentieth century. The greatest advance in adhesives, the development of adhesives based on synthetic rubbers, on other vinyl-type polymers, and on the epoxy resins, has occurred within the past twenty-five years.

Animal Glue (12). The most important of the protein adhesives is obtained from cattle and other animal hide and bones by hydrolysis of the collagen (qv). Hide glues are higher in molecular weight than bone glues and therefore stronger. After a pretreatment, the hides are treated with mineral acid and cooked to extract the solubilized collagen. The liquors are filtered, evaporated, dried, and ground.

Bone glues are made primarily from fresh or "green" bones. These bones are subjected to cycles of steam under pressure and extraction with hot water. The glue liquors obtained are filtered, centrifuged to remove grease or fat, evaporated, dried, and ground.

An important property of the animal glues is their ability to gel. The *jelly strength*, measured in Bloom grams, varies from 50 to 200 g for bone glues, and from 50 to 512 g for hide glues. Price increases with the jelly value. Glues are classified, according to test grade, as follows: low, 10–149 g; medium, 150–266 g; medium high, 267–330 g; and high, 331–529 g. Jelly strength, determined by measuring viscosities of 12.5% solutions of glues, may vary from 42 to 191 mP at 25°C. The *jelly* (gel) *value* is obtained by gelling at 10°C and measuring the force required to depress the surface of the gel 4 mm with a plunger 0.5 in. in diameter.

Almost half of the animal glue produced goes into remoistenable *gummed tapes*. The greatest demand is for green-bone glues of low-test grade. Dextrin is replacing animal glue as the main ingredient of cheaper gummed papers, but glue is preferred for the better products. A typical bone-glue formulation contains 50% solids. Dextrin may be added to the extent of 10–20% of the protein to reduce cost and improve the wetting and tack characteristics. Other ingredients include wetting agents, plasticizers, and gel depressants. Salts and urea, for example, help the water remoisten the glue more quickly and make it stay tacky longer.

Gummed paper has a tendency to curl as the coating dries. To make labels that stay flat, it is necessary to break up the continuous film of adhesive. This can be accomplished mechanically by pulling the adhesive-coated paper over sharp edges in more than one direction. Alternatively, the glue–dextrin adhesive can be dispersed in an organic solvent, eg, toluene, so that it is deposited upon the paper or thin-film substrate in discrete droplets rather than as a continuous layer.

Animal glues are still preferred by the craftsmen for joining furniture. Glues of 251–315-g strength are employed for high-speed assembly gluing, which requires a short setting time. For slower assembly operations, as in veneering, the animal glues of 135–225-g strength are adequate. The rate of gelation can be decreased by incorporation of 3–10% of thiourea in the formulation.

Sandpaper and some types of emery cloth use animal glue or high-test grades to bind the silicon carbide, aluminum oxide, flint, garnet, emery, or other abrasive rings (see ABRASIVES). Hide glues are used. Both the "make coat," in which the abrasive grains are imbedded, and the top "size coat," which locks the grains, may be of animal glue. It has been found, however, that greater durability and heat resistance result when at least the size coat is a phenol–formaldehyde resin (see below, Phenolic Resins).

Other uses for animal glue include bookbinding, paper sizing and coating, textile sizing, the binding of cork granules, and the binding of compositions for match heads.

Fish Glue (13). This adhesive is derived from fish skins, especially those of the cod. The skins are washed in cold water to remove salt, then cooked to extract the 5–7% glue solids. This liquor is concentrated to 40–50% solids, and a bactericide is added.

Fish glue is a weaker adhesive than animal glue, but it is adequate by itself for many uses as a household adhesive. It is a useful modifier for animal glue in the manufacture of gummed tape. Because of its low molecular weight, the addition of 10% of fish glue improves remoistenability, even in cold water. Fish glue is sometimes incorporated into dextrin formulations for envelope seals and seams, to give better tack. It is also used to some extent with poly(vinyl acetate) and rubber latexes.

Casein (qv) (14). The chief protein of milk, casein, is precipitated from skimmed milk by acidifying the milk to pH 4.5, using either hydrochloric acid or lactic acid produced by fermentation of the lactose in the milk. The curd is washed, dried, and ground. Yield is approximately 3%. Most of the casein used in the U.S. is imported from Argentina.

Casein is soluble at alkaline pH values. To make a casein adhesive joint resistant to water it must be insolubilized. The two principal methods for accomplishing this involve either lime or formaldehyde donors. The lime functions via the free carboxyl groups present in the protein, forming a gel of calcium caseinate.

One type of dry prepared resin glue contains casein, lime in excess, and a sodium salt, such as sodium phosphate, sulfite, carbonate, and/or fluoride. When the glue is dispersed in water, the sodium salt raises the pH, causing the protein to dissolve. After the joint has been made, the excess lime causes the casein to gel. Lime, in excess, results in faster gelation and better water resistance.

Alternatively, water resistance is attained through the use of formaldehyde donors, such as urea–formaldehyde resin or hexamethylenetetramine. Formaldehyde and methylol groups react with the amine groups abundantly available in the casein molecule and make the resulting product water-soluble.

Casein glues may be compounded with thickeners, such as alum or sodium sulfate; thinners, such as sodium sulfite or sugar; inert organic fillers; nondrying oils; preservatives; and humectants, such as glycerol or sorbitol. Casein may be used alone or blended with other adhesive materials, such as sodium silicate, soybean meal, blood albumin, natural rubber latex, and neoprene.

Wood is the principal substrate bonded with casein. Casein is the favored adhesive for structural laminates that are to be used indoors. These are composites of sawn wood, glued with the grain parallel. Huge arches and beams, laminated with casein, function as supporting members for roofs in churches, halls, and barns. In woodworking, casein adhesives are cheaper than animal glue. In structural plywood, they are less expensive than phenolic resin, but they are suitable only for interior-grade plywood that is not to be exposed to high humidity.

Ammoniacal casein solutions are blended with styrene–butadiene (SBR) or neoprene latexes to give adhesives suitable for bonding aluminum foil to paper. For label-making uses, such as for beer and soft-drink bottles, a properly formulated casein provides water resistance superior to adhesives that are obtainable with dextrin.

In addition to the uses enumerated above, casein has a variety of miscellaneous uses as a bonding agent of paper or wood products.

Soybean and Blood Glues (15).   These glues are economical and effective components of adhesives for plywood; they cause rapid gelation. They may be used as the main polymeric component, incorporated as extenders for phenolic resins, or blended with casein or sodium silicate.

Soybean flour contains both protein and carbohydrate. For adhesives, the flour is generally dispersed in aqueous sodium hydroxide. Along with sodium hydroxide, other alkaline bivalent metallic ions are incorporated, such as calcium hydroxide, to lengthen the "open time" (working time), and to insolubilize the proteinates, thus improving the water resistance of the adhesive joint. The disadvantage often noted is that an alkali causes a characteristic stain in the wood.

One-package soybean-glue formulations are dry powders containing sodium salts and calcium hydroxide that give a highly alkaline pH upon addition of water.

In addition to the alkaline earths, many polyfunctional materials are used as crosslinking agents for the dispersed soybean proteins. Typical denaturants and cross-linking agents include sulfur compounds, such as carbon disulfide; soluble metal salts; epoxides; and formaldehyde donors, such as dimethylolurea, hexamethylenetetra-mine, and trimethylolphenol. Small proportions, usually under 1% of the dry weight of soybean flour, are sufficient. The soybean protein is protected from fungal attack also by pentachlorophenol or other wood preservative. Fillers such as wood flour, walnut-shell flour, and clay result in lower cost, but they also lower performance of the adhesive.

Blood glues are preferably made from soluble dried beef blood, a by-product of meat-packing operations. They are more easily dispersed in alkali and denatured by formaldehyde than the soybean glues. The preferred antifungal agent, which also provides additional strength, is phenol–formaldehyde resin. In phenolic resins for outdoor plywood, a concentration of blood up to 10% is permitted by the Douglas Fir Plywood Association (Commercial Standard CS 45–60). Compositions containing larger proportions of blood are limited to indoor applications.

Both blood glues and soybean glues can be safely utilized in the cold for the laminating of wood. However, for rapid production of glued assemblies, with several cure cycles performed per hour, it is necessary to employ temperatures as high as 280°F, with pressures of 175–200 psi applied to the joint.

Polyamides (16).   There are two main categories of polyamides in adhesives. Most important at present are those derived from "dimer acids," but the nylons are among the promising new components of adhesive compositions.

"Dimer acids" are polymerized fatty acids, usually containing 36 carbon atoms and 2 carboxyl groups. Some trimer, higher polymers, and monomer are also present. The acids are condensed with polyamines, eg, ethylenediamine and diethylenetri-amine. The largely aliphatic structure results in products having high elongation properties.

Hot melts, made by the condensation of a dicarboxylic acid with an equivalent amount of a diamine, are thermoplastic polyamides that are essentially neutral. These resins form the basis for hot-melt or heat-seal adhesive compositions. They are used in packaging, for the bonding of polyethylene films for label papers and glassine, and other paper-bonding applications. They are compounded with minor amounts of rosin esters for increased tack, with paraffin for resistance to blocking, and with plasticizers for increased fluidity and low-temperature flexibility.

The products of condensation of "dimer acid" with an excess of polyamine are also termed polyamides, but their chief utility comes from the presence of free amine groups. These materials are useful as curing agents for epoxy resins to impart flexibility and toughness to the adhesive.

Starches and Dextrins (17). These carbohydrates are the adhesives that are used in largest tonnage. They are readily available, low in cost, and easy to apply from water dispersions. They are the principal adhesives for the bonding of paper products. Poly(vinyl acetate) and other synthetic adhesive bases, which give superior bond strength, tack, or other properties, are growing competitors for the paper-bonding market.

Starch is a polymer of glucose containing both *amylopectin* and *amylose* structures. The amylopectin molecules are branched, generally longer, amorphous, and more readily soluble. The amylose grouping is straight-chain, crystalline, and not readily soluble. The most widely used starches for adhesives are derived from corn, tapioca, or potatoes. They have a predominantly amylopectin structure; but they contain sufficient amylose to give them the property of gelling or "setting back" needed for best adhesive uses.

Unmodified starch can be dispersed in water by cooking with live steam; but the viscosity becomes excessive if other than a rather low concentration of starch is used. By suitable modification, however, starch polymers of greater solubility and viscosity control may be made. *Thin-boiling* or *high-fluidity starches* are made by acid treatment of starch slurries. The acid is left in the starch when it is dried. *Oxidized* or *chlorinated starches* are the products of aqueous sodium hypochlorite treatment of starch. Because of their low color, they are in demand for the sizing and coating of printing papers.

The *dextrins* are the outstanding starch derivatives made for adhesive purposes. Thorough treatment with heat and acid, the starch molecules are hydrolyzed into small fragments, then repolymerized into highly branched, readily soluble polymer molecules of moderate size. The *British gums* are dextrins made with heat as the principal agent. They are the highest in molecular weight and are the strongest adhesives among the dextrins, but maximum usable solids content is only approximately 25%. *Canary dextrins*, made by an acid treatment, are the lowest in molecular weight; materials of this type are available that give a viscosity of only 1700 cSt at 60% concentration. The *white dextrins* have the lowest color of the three. Their molecular weight is intermediate between those of the other two.

Most *corrugated boxboard* for making cartons is bonded with starch. A fraction of the starch needed to formulate the adhesive is gelatinized with aqueous caustic. This is blended with a concentrated suspension of unmodified starch granules. A typical starch adhesive formulation also includes bentonite as a thixotroping agent, borax to speed gelatinization, and a small quantity of formaldehyde as the mold inhibitor and crosslinking agent. The paste is applied cold to the corrugated flutes and to the liners. Upon subsequent exposure to heat, the starch granules swell and burst, forming a strong bond. This ingenious blend of ungelatinized and gelatinized starches permits high solids content and good strength.

Paper bags are joined at the seam with thin-boiling starches or dextrins, often compounded with urea–formaldehyde resin for superior water resistance. Envelope seals and other gummed papers for postage stamps, labels, etc, make use of the lower-molecular-weight dextrins containing 55–65% solids. Plasticizers and stabilizers are also added. For paper-box manufacture, dextrins are often compounded with borax as an aid to rapid gelatinization. Carton sealing, case sealing, and tube winding are other major uses for starch adhesives.

Good adhesion at low cost is an important property of starch in many other uses, such as in the sizing and clay coating of paper, and the sizing of textile yarns prior to weaving.

Cellulose Derivatives (18). The oldest of the man-made adhesives, *cellulose nitrate* (nitrocellulose, pyroxylin), is still utilized in some household cements. In a typical formulation, the polymer containing 11.4% nitrogen is plasticized with camphor and dissolved in an alcohol–ester–ketone mixture. Other cellulose nitrate formulations, containing wax, go into the heat-seal adhesive coatings on cellophane. Cellulose nitrate gives a very strong bond to polar substrates, but it is too inflammable to be suitable for many industrial applications.

Ethylcellulose and *cellulose acetate-butyrate* are among the polymers that have been employed as bases for hot-melt compositions. For such applications, they are heavily formulated with waxes and plasticizers.

Rubber-Based Adhesives (19,20). These are the largest group of adhesive materials in terms of dollar value. The rubbers are used both as latexes and as solvent cements. Latex adhesives, with their low viscosity and high solids content, are advantageous in the manufacture of shoes, books, and textile laminates. Tire cord is bonded to rubber with mixtures of rubber latexes and resorcinol–formaldehyde resin. Most rubber for adhesives is utilized in solvent-cement form. The rubber is dissolved in the solvent by high-speed mixers. Other ingredients may include tackifiers, resins, fillers, softeners, antioxidants, vulcanizing agents, and sequestering agents.

For structural bonding applications, considerable success has been achieved with blends of the more polar synthetic rubbers mixed with phenolic resins. The resin provides high shear strength and heat resistance, whereas the elastomer is responsible for high elongation and high peel strength of the bond.

The tackifier is often the critical component. The addition of a tackifier characteristically increases the tack up to a certain point, beyond which the tack strength falls away abruptly. For many rubber–resin combinations, more resin than rubber is present in the optimum composition. The precipitous loss of tack strength beyond the peak is ascribed to phase inversion (19). Among the tackifiers in common use are partially hydrogenated rosin and rosin esters, polyterpenes, coumarone–indene resins (qv), low-molecular-weight styrene polymers, oil-soluble phenolic resins, and low-molecular-weight petroleum resins.

Fillers for rubber adhesives include carbon black, zinc oxide, clays, hydrated alumina, calcium carbonate, silicates, and others. Prominent among the softeners used are mineral oil and lanolin. Among the leading antioxidants (qv) are the aromatic amines, which are lowest in cost, and the substituted phenols, which are nonstaining and light in color. Several hindered phenols have received the approval of the Food and Drug Administration.

Major markets for rubber cements are for pressure-sensitive tape, and in the automobile, building, canning, shoe, and packaging industries.

Natural rubber (21) accounts for approximately one-half the rubber used in adhesives in the United States. Although more expensive than styrene–butadiene rubber (SBR), it has far better adhesive properties, particularly tack. Thus it is possible to coat two adherend surfaces with rubber latex or cement, let the water or solvent evaporate, and then bond the surfaces to each other with only the contact pressure needed.

In addition to tack, natural rubber has outstanding resilience, far superior to SBR and other synthetics. (The new all-cis synthetic elastomers have not yet had an influence upon adhesives technology.) On the other hand, natural rubber is more sensitive to heat and oxidation, it has poor creep characteristics unless highly vulcanized and it is more readily attacked by hydrocarbon solvents.

Natural rubber latex contains 35% or more of rubber. Compounding ingredients must often be emulsified before addition, to avoid coagulating the rubber. Solvents, added in minor proportions, cause the latex to become more sticky, almost to the point of coagulation.

Cements have as their essential ingredients smoked rubber sheet and petroleum solvents, eg, hexane, heptane, mineral spirits, gasoline, and toluene. In addition, they may contain such ingredients as calcium carbonate, zinc oxide, tackifiers, antioxidants, and curing agents.

Natural-rubber cements are used for pressure-sensitive adhesives, as well as for the bonding of shoes, cloth, upholstery, floor tile, wall tile, etc.

Reclaimed rubber (21) is a development of the past thirty years. Reclaimed rubber adhesives are almost as good as natural-rubber materials, and are actually superior in adhesion when used to bond metals. In addition, they are much cheaper and can be applied more quickly because of the higher solids content that is feasible. Reclaimed rubbers are necessarily dark in color. This is not important in many applications but does preclude their use for some upholstery-bonding uses.

Reclaimed rubber cements and dispersions when mixed with asphalt have uses in automobiles as body sealers and sound deadeners, for attaching fabric to metal, for fastening floor mats to metal floors, and for bonding flexible trim to metal. A typical formulation contains 175 parts of asphalt emulsion to 100 parts of black reclaimed rubber, along with smaller proportions of clay, ester gum, emulsifier, and caustic. Reclaimed rubber mastics used in home construction to adhere plastic tile to wood subflooring or to concrete is another major use.

Styrene–butadiene rubber (22) (SBR) (see Butadiene polymers) does not have the good adhesive properties of natural rubber, nitrile rubber, or neoprene. However, it is lower in cost than these, and it has better heat-aging resistance properties than natural rubber. Furthermore, SBR absorbs less water, and the reinforced compositions have better strength retention. Their tendency to stiffen upon prolonged heat exposure can be an advantage in many adhesive applications.

SBR used in adhesives in cements is usually made primarily for other purposes. The styrene and butadiene are polymerized in emulsion in the presence of fatty acid and/or rosin soaps, using redox initiators, mercaptan chain-transfer agents, etc. Aromatic amines or phenolic materials such as hydroquinone are present as "short-stops" and antioxidants. Polymerization may be conducted at either 120–130°F to produce "hot" rubbers, or at 40–43°F for the "cold" rubbers. The hot rubbers are preferred for adhesives. SBR 1011 is widely used in pressure-sensitive tapes. SBR 1006 and 1012, with Mooney viscosity values of 50–58 and 95–115, respectively, are also useful. They are available in crumb form and may be dissolved without milling.

Styrene–butadiene rubber is not processed into adhesive formulations as readily as natural rubber, and does not have as good tack. Consequently higher proportions of tackifiers and plasticizers may be needed to make them into satisfactory adhesives. SBR can be processed in intensive mixers or on rubber mills. A typical laminating adhesive for cellophane is given in Ref. 22a: SBR, 9–15 parts by weight; coumarone–indene resin, 37–54 parts; and a mixture of microcrystalline wax and paraffin wax, 37–53 parts.

Other formulations omit the waxes but often incorporate fillers, such as clay and zinc oxide, vulcanizing agents, such as sulfur and accelerator, titanium dioxide pigment or carbon black as reinforcing pigments, antioxidants, etc. Oil-soluble phenolic resins may be incorporated as tackifiers, which are cured by heat. Aliphatic and aromatic hydrocarbons are among the most common solvents.

SBR solvent cements are used primarily for laminants, pressure-sensitive mass-coats, and for the splicing of tire treads. SBR latexes are used to bond fabrics to themselves as well as to rubber, paper, film, leather, and wood.

Nitrile rubbers (23) (termed "Buna N" in Germany) are copolymers of butadiene with acrylonitrile (see BUTADIENE POLYMERS). As the proportion of acrylonitrile is increased, the polymer acquires resistance to oil, increased solubility in ketones, esters, and aromatic and chlorinated hydrocarbons, and enhanced adhesion to polar substrates. Copolymers containing 25–45% acrylonitrile are suitable for adhesives; those in widest use are the copolymers containing 38% acrylonitrile, especially those with high Mooney viscosity values.

To obtain high strength for heavy-duty applications, nitrile rubber adhesives are blended with oil-soluble phenolic resins. Such a formulation may contain: 100 parts nitrile rubber, 38% acrylonitrile; 75–200 parts phenolic resin, novolac type; 5 parts zinc oxide; 1–3 parts sulfur; 0.5–1 part accelerator; 0–5 parts antioxidant; 0–1 part stearic acid; 0–20 parts carbon black; 0–100 parts filler; 0–10 parts plasticizer; methyl ethyl ketones, qs.

Compositions like this are applied to tapes and fabrics of cotton, rayon, nylon, or glass cloth, and dried. These tapes are utilized in the aircraft industry for the bonding of aluminum facings to honeycomb cores. The adhesive joints made with heat and pressure have high strength, and resistance to elevated temperature and to fatigue. For still greater temperature resistance, part or all of the novolac may be replaced by resole-type phenolic resins.

Nitrile rubber–phenolic resin adhesives are also suitable for bonding phenolic plastics. Without phenolic resin in the formulation, compositions containing nitrile rubbers are excellent adhesives for vinyl chloride polymers. The nitrile rubbers and poly(vinyl chloride) are nearly equal in their solubility parameters (approx 9.5); hence they are about as compatible as can be expected from polymers of differing chemical compositions. The nitrile rubbers may be blended with vinyl chloride copolymers for enhanced adhesion to vinyl sheeting, leather, and fabrics. Naturally, nitrile rubber cements are excellent adhesives for bonding cured or uncured nitrile rubbers to themselves.

Neoprene (24) has become the generic term for 2-chlorobutadiene polymers (qv).

Although most neoprene adhesives are used as solvent cements, the latexes are now finding increased usage. Most commercial grades are copolymers in which a small proportion of another monomer is present. Recently a neoprene has become available with substantial amounts of acrylonitrile as the comonomer. Known in the trade as neoprene ILA, it is an especially valuable raw material for adhesives because it possesses the specific adhesion characteristics of the nitrile rubbers as well as the high cohesive strength and contact bonding properties of the neoprenes.

Neoprene cements may be designed for curing or noncuring properties. Curing cements are formulated with crosslinking agents such as *s*-diphenylthiourea, litharge (lead oxide), and/or aliphatic polyamines to effect crosslinking at either elevated or room temperature. Magnesium oxide and zinc oxide are also important components of neoprene cements. These metal oxides are needed to neutralize the hydrochloric acid gradually liberated by the polymer. Along with antioxidants, they are responsible for the excellent resistance of neoprene films to ozone and oxidation. A basic formula is: neoprene, 10 parts; resin, 1–5 parts; magnesia, 4 parts; antioxidant, 2 parts; zinc oxide, 5 parts; and solvent, qs.

The most common solvents are aromatic or chlorinated hydrocarbons, methyl ethyl ketone, and blends of these materials with aliphatic hydrocarbons. In some cases a pair of solvents, neither of which is a solvent for neoprene when used alone, is suitable when combined in the right proportion, eg, 50:50 mixtures of acetone and hexane, or ethyl acetate and hexane.

The shoe industry is the largest outlet for the neoprene cements. Shoe adhesives may be either temporary or permanent. A temporary adhesive is used, for example, in the laying on of soles preparatory to nailing or sewing. There is a trend in the shoe industry toward elimination of nails and stitching through the use of permanent neoprene adhesives.

Neoprene is the basic material in the contact cements used in installing laminated plastic counter tops in kitchens and bars. The cement is applied to both substrates, the solvent is allowed to evaporate, and the faying surfaces are carefully juxtaposed. Because of the exceptionally high tack of the neoprenes, it is difficult to correct a misalignment once the two prepared surfaces are in contact.

Like the nitrile rubbers, the neoprenes are often blended with phenolic resins to attain greater strength and heat resistance. The phenolics used are preferably of the oil-soluble, heat-reactive type. These resins react with the magnesium oxide to give enhanced cohesive strength and heat resistance. They are useful for the bonding together of glass, rubbers, and metals.

Isobutylene polymers (25) have better aging characteristics than the types of rubber discussed heretofore by virtue of their chemical composition (see BUTYLENE POLYMERS). Polyisobutylene is completely saturated, whereas the butyl rubbers contain only a small amount of unsaturation (1–3% isoprene). Since isobutylene is polymerized with Lewis acid catalysts in the presence of only trace quantities of water, the polymers are free from the emulsifiers present in latex polymers.

The polyisobutylenes and butyl rubber are soluble in aliphatic, aromatic, and chlorinated hydrocarbons.

Polyisobutylene is a useful component of pressure-sensitive adhesives. A suggested adhesive for surgical tape contains: 100 parts polyisobutylene; 50 parts zinc oxide; 50 parts hydrated alumina; 50 parts mineral oil; 70 parts phenolic resin; and 600 parts solvent naphtha or heptane.

Because of their excellent aging characteristics, the isobutylene polymers are being used more and more in calking compositions (see SEALANTS) and sealers for curtain-wall construction operations. For this purpose, they are blended with such other ingredients as asbestos fiber, calcium carbonate, graphite, bentonite, and rosin derivatives. A minor proportion of mineral spirits or kerosene is added to aid spreading.

Polysulfide polymers (qv) (26) are at present the leading materials in use as sealants for curtain-wall construction. These completely saturated polymers are viscous liquids that are compounded with carbon black and cured with lead dioxide pastes to yield elastomers having long-term weather resistance. For longer pot life, cure is retarded by the addition of such materials as stearic acid, aluminum distearate, or lead stearate.

It is advisable to apply primers to the substrates before application of the polysulfide compositions. For glass, aluminum, and steel, useful primers have been based on neoprene, chlorinated rubber, low-viscosity polysulfide, and furan resin. The adhesion of the polysulfide can be enhanced by incorporation of phenolic resins.

Sealants are derived from polysulfide polymers having molecular weights of approximately 4000 and viscosities of approximately 4000 cP. Polymers of lower molecular weight and viscosity (approx 1000 and 1000 cP, respectively) are incorporated into epoxy resin compositions to provide flexibility.

Phenolic Resins (qv) (27). These resins are the most important of the synthetic adhesives. They are the preferred binders for such products as outdoor-grade plywood, electrical laminates, abrasive wheels, brake linings, and glass-wool insulation. In conjunction with neoprene and especially nitrile rubber, they are used for the structural bonding of metals to provide high tensile and peel strength, resistance to elevated temperatures, and shock resistance.

Phenolic resins are the condensation products of phenol and formaldehyde. In *resoles*, or one-stage phenolics, the ratio of phenol to formaldehyde is from 1:1 to 1:1.5. In the presence of alkaline catalysts, crosslinking is brought about at elevated temperatures. Adhesives for plywood are essentially solutions of low-molecular-weight condensation products of phenol and formaldehyde in aqueous sodium hydroxide. Acid catalysts may also be used in one-stage phenolics and give room-temperature cures, but they cause degradation of wood and paper.

In *novolacs*, or two-stage phenolics, the ratio of phenol to formaldehyde is from 1:0.8 to 1:1. The reactants are condensed in the presence of acid catalysts, eg, oxalic acid. These resins do not crosslink unless additional formaldehyde is supplied, eg, by addition of hexamethylenetetramine ("hexa" or HMTA). Upon thermal decomposition, the latter also releases ammonia, which functions as a catalyst to promote cure.

The soluble novolac-forming intermediates are chemically stable, both in organic solution and in dry form. Powders are convenient for making grinding wheels and other assemblies in which drying might be a problem. The novolacs are more expensive than resoles.

Phenol–formaldehyde resin is available in the form of a glue film, carried on tissue paper. The resin is still soluble and contains alkaline catalysts. Although expensive, this form of film adhesive can be useful for faying of very thin or highly porous veneers.

The Douglas-fir plywood industry utilizes phenolic resins for the exterior grades.

These must be capable of strongly resisting delamination during and after immersion in boiling water. The phenolic adhesive may be combined with up to 10% of soluble blood, as well as fillers, such as oat-hull lignins, bark, walnut-shell flour, wood flour, and wheat flour. A typical phenolic adhesive formulation for exterior-grade plywood contains: phenol–formaldehyde (formaldehyde in excess), 25–30%; blood solubles, 0–10%; caustic and soda ash, 2–4%; filler, 10–25%; and water 31–63%. The water content is varied depending on the substrate and the required pot life. Cure is effected in 5–15 min at elevated temperature and pressure in large multiple-opening platen presses. The shorter press times are achieved through higher formaldehyde:phenol ratios and greater catalyst concentrations. The presence of blood also speeds the cure time.

A widely used filler–extender is oat-hull lignins, but bark is preferred by some plywood makers because its phenolic content permits an even greater reduction in the proportion of synthetic resin.

Paper laminates for electrical insulation purposes are made by impregnating webs of paper, cloth, etc, with phenolic resin, drying, assembling, and curing under heat and pressure. For counter tops and table tops, the brown color of cured phenolic resins is objectionable. They are outer-faced with decorative layers impregnated with melamine resins to overcome this objection.

In making coated and bonded abrasives, phenolic adhesives have largely displaced animal glue and shellac as binders for the abrasive grains, both in the "make coat," in which the abrasive is embedded, and in the protective "size coat."

Resorcinol Resins (28). The resins of resorcinol are far more expensive than their phenolic prototypes, but are useful where a room-temperature cure is necessary. Two types are common: resorcinol and phenol–resorcinol. The unmodified resorcinol resins are prepared by partial reaction of resorcinol with formaldehyde in molar ratio of from 1:0.6 to 1:0.65 in the absence of a catalyst. The condensate is diluted to 65% concentration with ethanol. This solution is mixed just before use with a filler, such as walnut-shell flour, and a source of formaldehyde, such as paraformaldehyde. The glue is applied to the wood surfaces, most of the alcohol is allowed to evaporate, and then the adherends are brought together. Cure takes place in 15–45 min, depending upon temperature. Although the wood pieces must be held together by jigs or clamps, no oven or hot press is required. Consequently, the resorcinol adhesives are convenient for the lamination of lumber to form arches, boat keels, and other massive wood structures that may be exposed to water or high humidity.

In the manufacture of tires, the reinforcing cord is bonded to the rubber carcass with "RFL" adhesives; these are mixtures of resorcinol–formaldehyde resin and rubber latexes. A terpolymer formed of vinylpyridine, styrene, and butadiene monomers is the preferred latex for nylon cord. For rayon, the same polymer is admixed in minor proportion with a styrene–butadiene polymer latex, which is less expensive.

Amino Resins (qv) (29). These include urea–formaldehyde, melamine–formaldehyde, and various modifications thereof. The liquid urea resin adhesives are made by condensation in water of urea with 1.5–2 moles of formaldehyde. The resin is then stored while alkaline. Before use the resin is blended with fillers such as wood flour, shell flours, and cereal flours, acid catalysts such as ammonium chloride, and sometimes with buffering agents to prolong the working life. Powdered or "neat" resins are made by spray drying. More expensive than others, their use is restricted largely to "do-it-yourself" applications.

The largest application for amino resins is in the manufacture of interior-grade plywood, primarily hardwood. With sufficient catalyst, cure can be effected at room temperature under pressures of 150–200 psi. For assembly wood gluing, furfuryl alcohol may be incorporated for improved gap filling and resistance to crazing. Cold-press ureas require several hours for cure. By pressing at 240–260°F, cure time can be shortened to 1–15 min. depending on the catalyst system.

Melamine–formaldehyde resins are made by the reaction of melamine with 3 moles of formaldehyde at slightly alkaline pH; they are then stabilized with strong alkali. Far more expensive than the urea resins, they are employed where superior resistance to water and elevated temperatures are required. For economy they are often diluted with urea resins.

Like the ureas, the melamine resins are substantially free from color. They are used in the decorative upper layers of counter tops and in table tops having cores of phenolic laminate, particle board, or hardboard.

Epoxy Resins (qv) (30). The epoxy resins used for adhesives are most commonly the condensation products of epichlorohydrin and bisphenol A (4,4'-isopropylidene-diphenol). The viscous liquid epoxy resins that have epoxide equivalent weights of 170–300 are particularly useful. They are cured with reactive hardeners and/or catalytic hardeners, which are added just before application of the adhesive. Many of the hardeners function well at room temperature. Cure takes place without the production of volatile by-products that occurs with the phenolic and urea resins; consequently, the joint may be made with only the minimum pressure necessary to hold the pieces together. Adhesion to metals and glass is excellent, and shrinkage is low. The adhesive joints show good resistance to fatigue, creep, heat, moisture, and solvents.

Reactive hardeners include polyamines such as diethylenetriamine, triethylene-tetramine (TETA), *m*-phenylenediamine, methylenedianiline, and diamino diphenyl sulfone. The aliphatic amines function as curing agents at room temperature; the aromatic amines require elevated temperatures but produce products of higher heat resistance.

The reactive hardeners are used in substantially stoichiometric proportion, eg: epoxy resin epoxide equivalent 190, 100 parts; and triethylenetetramine, 11 parts.

On the other hand, tertiary amines such as tris(dimethylaminomethyl)phenol are incorporated in lower proportion, eg, 5%, because they catalyze the conversion of the epoxy intermediates to polyethers. The boron trifluoride–ethylamine adduct may also be used for this purpose at elevated temperature.

Anhydrides, such as phthalic anhydride or chlorendic anhydride, have both a re-active and catalytic action with epoxy resins; they give best results when used in a ratio of slightly less than one anhydride group per epoxide group. Like the aromatic amines, these aromatic or condensed ring anhydrides give cured adhesive-bonded products of superior heat resistance. For still higher heat resistance, resins are em-ployed having more than two epoxide groups, such as the polyglycidyl ethers of novolacs and of other polyphenols. Blends of epoxy and phenolic resins combine the superior adhesive qualities of the epoxies and the superior strength retention at elevated temperatures of the phenolics.

Drawbacks to the use of epoxy resins are their high cost, the danger of dermatitis from handling many of the amines, and the inconvenience of two-part systems (the resin and the hardener are kept separate until use). That these difficulties are surmountable is indicated by the rapid surge in epoxy-resin usage. Cost can often be reduced though incorporation of 50–500 phr of fillers such as silica, aluminum powder, or iron oxide. Dispensing equipment is available that minimizes handling problems.

For increased adhesive toughness, the epoxy resin is blended with liquid polysulfide resins of molecular weight 1000, or with polyamide–amines derived from "dimer acid" (see above); these latter function as reactive hardeners. Very high peel strength can be obtained through addition of nylons.

Among the more important uses of epoxies are the bonding of aluminum in aircraft structures, and the patching of steel automobile fenders, of wooden or polyester boats, and of concrete structures.

Poly(vinyl Acetate). Poly(vinyl acetate) and related polymers (31) are the fastest growing of the thermoplastic adhesives. They are used principally for the bonding of paper, especially in packaging applications.

Most of the vinyl acetate resins are prepared by emulsion polymerization. The latexes commonly contain 55% solids. In addition to the vinyl acetate polymer, protective colloids such as poly(vinyl alcohol) may be present in significant proportion. Other possible components of these adhesive formulations are starches and dextrins, rosin derivatives, plasticizers such as dibutyl phthalate, and small amounts of organic solvents and preservatives.

The homopolymers of vinyl acetate are adequate for such applications as pastes for paper bags, seam adhesives, pigment binders, household adhesives, and general adhesives for paperboard, wood, and textiles. Crotonic and other acid copolymers, redispersible or soluble in alkali, are used in bookbinding and bottle-label applications. Copolymers with maleic and acrylic esters have superior adhesion, flexibility, and film-forming characteristics.

Poly(vinyl alcohol) for adhesives application is usually made by hydrolysis of about 88% of the acetate groups in poly(vinyl acetate). It is water-soluble and therefore generally compatible with starches and dextrins. Vinyl acetate polymer latexes usually contain some poly(vinyl alcohol); but for envelopes, stamps, and gummed papers, the addition of considerably more poly(vinyl alcohol) results in better remoistenability, gloss, and resistance to blocking.

Inorganic Adhesives (32). Sodium silicate and other inorganic adhesives are in competition with the organic adhesive materials on the basis of both price and performance. The silicates, among the adhesive materials of lowest cost, contain ratios of alkali with silica of from 1:2 to 1:3.5, having pH values of 13–11, respectively. The more alkaline materials have better wetting power and lower viscosity, whereas the more siliceous materials dry more readily, giving adhesive joints that are less sensitive to water. Principal applications for the water-soluble silicate adhesives are in the manufacture of corrugated board, fiberboard, spiral wound tubes, cans, and drums.

Other inorganic cements are made from phosphates, zinc oxychloride, calcium silicate–alumina–iron oxide (portland cement), lime, and gypsum (plaster of paris).

Recent Developments

Although almost every synthetic and natural polymer has some conceivable potential as an adhesive, the following materials have already demonstrated their utility:

Isocyanates of molecular weight 200–1000, usually containing 2 or 3 isocyanate groups per molecule, have merit for the bonding of polyester tire cord to rubber, the bonding of plasticized vinyl compositions to fabric, and the laminating of polyurethan foam to fabric. The first two uses named, at least, are quite difficult to achieve with other adhesives. The isocyanates react readily with hydroxyls to form urethans, and with amines or water to form substituted ureas. See POLYURETHANS.

2-Cyanoacrylates (qv under ACRYLIC ESTER POLYMERS), perhaps the most expensive of commercially available adhesives, are fluid monomers that polymerize in seconds upon exposure to moisture. They give bonds of high strength with most polar substrates. They deteriorate upon prolonged exposure to high humidity.

Poly(vinyl ethers) make possible the formulation of pressure-sensitive compositions for "permanent" labels.

Acrylic ester polymers (qv) are used for the bonding of nonwoven fabrics, where they have demonstrated excellent light stability and resistance to drycleaning or laundering. They are also being investigated for hot melts and in pressure-sensitive compositions. Especially interesting are copolymers containing carboxyl, hydroxyl, or other functional groups that facilitate crosslinking of fibers in nonwoven fabrics (qv).

Silicones (qv) are generally *abhesive*, but are of interest for making pressure-sensitive tapes having moderate tack, even on such nonpolar substrates as polyethylene, polypropylene, and polytetrafluoroethylene. Copolymers containing functional groups are of great interest. Although expensive, the silicones are of interest for uses as curtain-wall sealants because of their excellent aging characteristics.

Polybenzimidazoles (33), synthesized from aromatic tetramines and diphenyl isophthalate, show short-term resistance to temperatures as high as 500°C, which is unusual for organic materials. They appear to have special promise in metal-bonded products in which heat resistance is critical.

Glass frits that are partly fused may provide bonds for metals that will be useful at even higher temperatures. The most likely compositions are those in which the metal adherends match closely the coefficient of thermale xpansion of the adhesive.

Selecting an Adhesive

Several factors must be taken into consideration in the selection of an adhesive: (*1*) The adhesive must wet the substrates. (*2*) If the substrates are impervious and nonabsorbent, the adhesive must be free from water or organic solvent. (*3*) Low cost may be needed, not only in the adhesive itself, but also in the method of application. (*4*) The adhesive should be no more rigid than the adherends; otherwise stresses will concentrate in the adhesive layer. (*5*) After setting, the adhesive joint must be able to withstand environmental conditions.

Resistance to high temperatures, to prolonged stress, to water, and to various chemicals, and low electrical conductivity may be required, depending on the intended use.

Table 3, compiled by Reinhart and Callomon (7) for a project sponsored by the Nonmetallic Materials Division, Air Force Materials Laboratory, Aeronautical Systems Division, Wright-Patterson Air Force Base, Ohio, indicates the preferred adhesive materials for use with various substrates.

Test Methods

Test methods (34) for adhesives are devised by ASTM Committee D 14. Adhe-

Table 3. Selecting Adhesives for Use with Various Substrates[a,b] (7)

	Leather	Paper	Wood	Felt	Fabrics	Vinyl plastics	Phenolic plastics	Rubber	Tile, etc	Masonite	Glass	Metals
metals	1, 4, 21, 24, 25	1, 21, 22	1, 4, 11, 13, 21, 31, 32, 33, 35, 36	1, 5, 22	1, 21, 22, 24	25, 36	3, 13, 21, 31, 32, 33, 35, 36	13, 21, 22, 31, 32, 33, 35, 36	5, 6, 13, 22, 35, 36	5, 6, 13, 22	13, 32, 33, 34, 35	11, 13, 31, 32, 33, 36
glass, ceramics	1, 4, 13, 24	1, 21, 22	1, 13, 21, 31, 32, 33, 35, 36	1, 5, 6, 21, 22	1, 21, 22, 24	25, 36	3, 13, 21, 31, 35, 36	21, 22, 31, 35, 36	4, 22		4, 13, 32, 35, 36	
tile, etc	1, 4, 21, 24	1, 21, 22	1, 5, 6, 21, 22	5, 6, 21, 22	5, 6, 21, 22, 24	25, 36	3, 13, 36	21, 22, 31, 35, 36	4, 5, 6, 22	5, 8, 13		
masonite	1, 21, 24	1, 21, 22	1, 5, 6, 21, 22	5, 6, 21, 22	5, 6, 21, 22, 24	25, 36	3, 13, 36	21, 22, 31, 35, 36	5, 6, 22			
rubber	21, 24	21, 22	21, 22, 33, 35, 36	21, 22	21, 22, 23	25, 36	21, 22, 36	21, 22, 31, 35, 36				
phenolic plastics	21, 24, 25	21, 22	11, 13, 21, 24, 32, 33, 36	21, 22, 25, 36	21, 22, 24, 25	36	13, 32, 33, 36					
vinyl plastics	21	21	21	21	21	25, 36						
fabrics	21, 22, 23, 24	21, 22, 23	21, 22, 23	5, 21, 22, 23	1, 21, 22, 23							
felt	21, 22, 23, 24	21, 22, 23	21, 22, 23	5, 22								
wood	21, 22, 23, 24	2, 21, 22	1, 11, 12, 14, 15, 36									
paper	21, 22, 23, 24	2, 4, 21										
leather	1, 4, 21, 22, 23, 24											

[a] Adhesive number code:

Thermoplastic
1. poly(vinyl acetate)
2. poly(vinyl alcohol)
3. acrylic polymer
4. cellulose nitrate
5. asphalt
6. oleoresin

Thermosetting
11. phenol–formaldehyde (phenolic)
12. resorcinol, phenol–resorcinol–formaldehyde
13. epoxy resin
14. urea–formaldehyde
15. melamine, melamine–urea–formaldehyde
16. alkyd resin

Elastomeric
21. natural rubber
22. reclaimed rubber
23. butadiene–styrene rubber (SBR)
24. neoprene
25. nitrile
26. silicone

Resin Blends
31. phenolic–vinyl
32. phenolic-poly(vinyl butyral)
33. phenolic-poly(vinyl formal)
34. phenolic–nylon
35. phenolic–neoprene
36. phenolic–butadiene–acrylonitrile rubber

[b] Courtesy Air Force Materials Laboratory, Aeronautical Systems Division, Wright–Patterson Air Force Base, Ohio.

sives are evaluated for viscosity (ASTM method D 553), consistency (D 1084), storage life (D 1337), pot life (D 1338), blocking (D 1146), penetration, and curing rate. It is proposed that tack be measured with the aid of a cantilever–bar device that pulls apart the freshly made joint, with separation force being measured as it occurs in the adhesive joint.

Tensile strength is measured by pulling a cured adhesive joint apart at right angles to the glue line (D 897, D 1344), or in shear (D 1002 for metal, D 906 for wood). Shear strength may also be determined by compression loading (D 905). Peel strength is determined by measuring the pulling force needed to separate adherends when force is applied to the glue line at an angle of 180°, if one adherend is flexible and the other is rigid (D 903). Another method (D 429) utilizes a 90° angle. Impact strength, a measure of the energy lost in breaking a specimen in a shear test with rapid loading, is determined with a pendulum–hammer device (D 950). Cleavage (D 1062), creep, fatigue (D 1002), and flexural loading strength (D 1184) are also of interest in determining and reporting qualities of materials and adhesives. A comprehensive monograph dealing with testing of adhesives in use has recently been published (35).

There are many ASTM tests for permanence of an adhesive joint, such as exposure to chemicals, light, heat, salt spray, etc. The supplier or end user may select from the standard tests, but may also wish to devise his own procedures for evaluation, by ad hoc methods devised to simulate performance of adherend products under intended usage conditions.

Bibliography

1. *ASTM Standard on Adhesives*, sponsored by Committee D 14, American Society for Testing and Materials, Philadelphia, 1961.
2. J. Delmonte, *The Technology of Adhesives*, Reinhold Publishing Corp., New York, 1947.
3. N. A. deBruyne and R. Houwink, eds., *Adhesion and Adhesives*, Elsevier Publishing Co., New York, 1951.
4. N. A. deBruyne, *Structural Adhesives*, Lange, Maxwell, Springer, London, 1952.
5. C. Leuttgen, *Die Technologie der Klebestoffe*, Wilhelm Pansegrau Verlag, Berlin-Wilmersdorf, 1953.
6. J. Clark, E. J. Rutzler, Jr., and R. L. Savage, *Adhesion and Adhesives, Fundamentals and Practice*, John Wiley & Sons, Inc., New York, 1954.
7. F. W. Reinhart and I. G. Callomon, "Survey of Adhesion and Adhesives," *WADC Technical Report 58-450* (1959).
8. J. J. Bikerman, *The Science of Adhesive Joints*, Academic Press, Inc., New York, 1960.
9. D. D. Eley, *Adhesion*, Oxford University Press, London, 1961.
10. I. Skeist, ed., *Handbook of Adhesives*, Reinhold Publishing Corp., New York, 1962.
11. I. Skeist, "Adhesives—An Economic Study," *Papers Am. Chem. Soc., Div. Chemical Marketing and Economics, Atlantic City, N.J., September, 1962*.
    Based on "A Study of the Market for Materials of Farm Origin in Adhesives Applications" carried out by Skeist Laboratories, Inc., under a contract with the U.S. Dept. of Agriculture.
12. J. R. Hubbard in Ref. 10, p. 114.
13. H. C. Walsh in Ref. 10, p. 126.
14. H. K. Salzburg in Ref. 10, p. 129.
15. A. L. Lambuth in Ref. 10, pp. 148, 158.
16. D. E. Floyd in Ref. 10, p. 425.
17. G. V. Caesar in Ref. 10, p. 170.
18. W. D. Paist in Ref. 10, p. 181.
19. F. H. Wetzel in Ref. 10, p. 188.
20. C. L. Weidner and G. J. Crocker, "Elastomeric Adhesion and Adhesives," *Rubber Chem. Technol.* **33,** 1323 (1960).

21. S. Palinchak and W. J. Yurgen in Ref. 10, p. 209.
22. J. F. Anderson and H. B. Brown in Ref. 10, p. 248.
22a. C. M. Carson (to Wingfoot Corp.), U.S. Pat. 2,541,689 (1951).
23. H. B. Brown and J. F. Anderson in Ref. 10, p. 229.
24. L. S. Bake in Ref. 10, p. 268.
25. K. F. Richards in Ref. 10, p. 221.
26. J. R. Panek in Ref. 10, p. 286.
27. C. B. Hemming in Ref. 10, p. 298.
28. C. B. Hemming in Ref. 10, p. 306.
29. J. F. Blais in Ref. 10, p. 310.
30. I. Skeist in Ref. 10, p. 323.
31. F. H. Norris and P. M. Draghetti in Ref. 10, p. 349.
32. J. H. Wills in Ref. 10, p. 83.
33. "Air Force Studies New Polymer Family," *Ind. Eng. Chem.* **55** (4), 12 (April 1963).
34. M. Petronio in Ref. 10, p. 61.
35. W. A. Neuss, ed., *Testing of Adhesives*, No. 26 in Tappi Monograph Series, Technical Association of the Pulp and Paper Industry, New York, 1963.

ACKNOWLEDGMENT. The author collected and analyzed much of the data on adhesive materials and uses given in this report during a contract research study of adhesives for the U.S. Department of Agriculture. Permission to publish, as a part of the Department's release of information from this study, has been granted to the author and to John Wiley & Sons, Inc.

Irving Skeist
Skeist Laboratories, Inc

GLUE, ANIMAL AND FISH

The term glue should be used to refer only to the adhesive material extracted from collagen, the primary constituent of all animal connective tissue abundantly found in skins, hides, sinews, and bones. The word is sometimes used more loosely to refer to all adhesive material irrespective of its origin; for this reason the expression *animal glue* can be used for greater precision. The proteins normally associated with hooves, horns, scales, feathers, and hair are keratin, not collagen, and cannot be used as sources for animal glue, even though statements have often been made to the contrary. Animal glue is closely related to gelatin (qv).

No written record is available to describe the first preparation and use of animal glue. It is logical to assume, however, that the cooking of animal or fish material led to the crude extraction of glue or gelatin from skin and bones and that the stickiness of this broth, as it approached dryness, eventually led to the discovery of its adhesive properties.

Animal glues are known by a variety of names which are usually descriptive of their source or appearance. Thus, *hide glue* may be called "skin glue," "chrome glue," "coney glue," or "technical gelatin." *Bone glue* may be described as "green bone glue," "extracted bone glue," or "dry bone glue." *Fish glues* are conventionally called "cold glues" or "isinglass" because of their tendency to remain fluid, and the light color and clarity of their films. All such terms have been used many years and may or may not be descriptive of the products produced today. For example, non-adhesive uses have given rise to the term "technical protein colloid," which is a general term applicable to all animal glue products when used because of their colloidal properties. Similarly specialty glue products may be described as flexible, liquid, water-resistant, iceproof, or painted glues (etc).

All of these are produced commercially for a number of uses and each may have its own trade name, which is usually descriptive of its purpose.

Manufacture of Hide Glues

The method of extracting hide glue is independent of the raw material, though the preparation of the raw stock prior to extraction may vary greatly.

Green stock is received in reasonably fresh unsalted condition and usually includes hide trimmings, splits, fleshings, tails, pates, lips, snouts, ears, and sinews (see also Leather). These are usually received separately and not as gross mixtures, since the processing is varied according to the physical form and composition of the raw stock. *Salted stock* or *dry stock* has been processed by salting or drying in order to preserve it while accumulating large quantities for shipment to the glue factory. Such stock may include any or all the material mentioned as green stock. *Chrome stock*, pieces or shavings, is received as a by-product from chrome tanning (see Leather). *Sheep stock* includes the pates primarily along with some sheepskin trimmings. *Horse stock, pigskin trimmings*, including ears and tails, and occasionally *pork feet*, are sometimes used as raw material for hide glue manufacture. Collagen from young animals produces a higher quality product than that from older animals and for this reason such raw material frequently commands a premium and is processed separately. *Calfskins* and *porkskins* are particularly suitable for gelatin manufacture and are not readily available for glue manufacture except under special conditions.

In the first step of the glue making process the stock is tumbled in large quantities of cold water to remove manure, dirt, blood, and salt until the wash water runs clear; the length of time depends upon the condition of the raw material. Dry and salted stock may be soaked prior to washing, since longer treatment is required to produce a soft saltfree raw material.

Curing is effected by treating the collagen raw material with either alkalies or acids. Which curing procedure is used depends upon the type of glue desired, the nature of the stock, and the facilities available. *Alkali curing* is usually done in vats or pits provided with some mechanical means for agitation. Lime is universally used because of its low cost, and because its low solubility (approx 2 g/liter) prevents excessive causticity. The pH of saturated lime water (12.0) is ideal for alkali curing. Care is taken to prevent packing of the raw stock which leads to heating and bacterial decomposition. It is often desirable to add a little caustic soda (or sodium carbonate, which generates caustic soda by reacting with the lime). During cold weather this is done because the cure is somewhat sluggish at low temperatures; in the summertime this procedure may accelerate penetration of the curing alkali and arrest bacterial decomposition.

The temperature, the nature of the raw material, and its physical form determine the selection of the curing formulation. In any case lime curing is time-consuming and requires from 3 to 10 weeks.

Certain types of raw material are received essentially cured. These are tannery limed fleshings, trimmings, or sheep stock from depilated pelts. Such material is completely cured in most instances and requires only back washing and neutralization to prepare for extraction.

Other forms of glue stock lend themselves to *acid curing* because of their physical form and composition. These are coney stock (the shredded rabbit skin by-product from the felting industry), bacon skins, fleshings, pork skin trimmings (when available), and pickled sheep skin trimmings. All of these cure readily in dilute acids (usually hydrochloric or sulfurous acid) and lend themselves to fast processing schedules since curing is complete in 6–16 hr.

The completion of the curing step is determined by visual examination of a cut cross section of the stock. Complete cure is evidenced by a uniform translucency throughout the swollen hydrated material. But complete penetration of the acid or alkaline curing agent is not sufficient in itself. Thus, it is not uncommon in the lime curing process for some excessive curing to occur on the exterior of thick stock before the interior is cured completely.

After curing, the stock is washed to remove excess curing agent, in the case of lime cure, until the wash water is clear. It is then washed with dilute sulfurous acid to solubilize the lime salts. The acid washing is continued until the outer surface of the stock has a pH of 3.   Water washing is then continued until the average pH of the entire stock is about 6.5.   The overall washing time requires from 8 to 12 hr depending upon the average thickness of the stock.

Acid-cured stock which has been treated with dilute hydrochloric acid or sulfurous acid (approx 0.5%) is washed with water until the average pH is between 3.5 and 4.0. This results in the extraction of a relatively acid liquor (pH 3.6–3.9).

Cooking schedules (*extraction* procedures) are planned with the general stock condition in mind.   All glue extraction methods include a series of hot water extractions, each extraction being made at a temperature about 10°F higher than the one before. Heavy stocks, which may be overcured on the surface are frequently given a hot wash to remove the easily extracted material so that it may be processed without further heating and hydrolysis.   Although curing prepares the collagen for conversion into glue or gelatin, it is the hot water, time, and temperature which effect the actual conversion of the insoluble collagen into glue or gelatin soluble in hot water.

The series of cooking operations is started by dropping the cured and washed stock into the extraction kettle containing hot water so that all the stock is curled by preheating.   This prevents packing and channeling.   The first run is then made by cooking at 140°F for about 3–4 hr.   It is drawn off through a false bottom containing a rough padding of excelsior (shredded wood) until any surface fat is about to run through. The amount of fat is held to a minimum by appropriate skimming during cooking.   The second run is made by just covering the stock with hot water at a temperature which will permit the second extraction to be made at 150°F.   Eight to twelve extractions are made in this manner until the solids extracted (less than 1%) are too low to justify concentration and further processing.   As the glue is extracted, the volume shrinks rapidly, and since each extraction just covers the stock, the volume of liquor decreases with each run.   The last runs are of low volume and are made just under the boiling point so as to ensure maximum extraction with minimum agitation and the highest possible liquor concentration.   All extraction methods employ water, heat, and time to convert the insoluble collagen into the soluble glue.   This primary hydrolysis is essential but subsequent hydrolytic breakdown of the glue is to be avoided.   The highest quality glue is extracted at the minimum temperature in the first runs, the lowest quality is obtained in the final runs.   Maximum solids are obtained in the middle runs, and thus, if one were to plot the product of glue yield and glue quality against the number of the extraction, a bell-shaped curve would be obtained.   Later extractions are subjected to almost 24 hr of cooking temperatures and maximum hydrolytic breakdown may be expected.   In the case of dry stock, 12–15 extractions may have to be made before satisfactory yields are obtained.   Dried raw material does not rehydrate readily and therefore requires more strenuous extraction schedules.   Salted stock poses no problem.

Throughout extraction and subsequent handling of the glue liquors, excessive agitation by pumping, churning, or splashing should be avoided in order to prevent the emulsification of fat.   Although tallow and grease are skimmed during extraction, an additional 2-hr settling period is customary in order to separate additional fat and to settle any particulate insoluble solids.   Extraction liquors vary in solids content from 1 to 6% and are called "light liquors" in contrast to the more viscous "heavy liquors," produced subsequently by evaporation.   Light liquors may be further "polished" by centrifuging or filtration if it is desired to remove even small amounts of fat and suspended matter.   Such steps are included when a technical gelatin is desired and a certain standard of clarity must be maintained. In this case acid curing is usually preferred because removal of fat, mucins, and other impurities, which cause turbidity, is much more easily effected at the lower pH.

The light-liquor stage is the critical point in glue manufacture because of the susceptibility to bacterial decomposition as a result of higher dilution and lower temperatures. Pasteurization temperatures of 140–150°F for the 2-hr settling period, or a reduction in pH to 5.0 or lower, negate any advantage in reduction of bacteria by increased hydrolysis.

The presence of thermophilic bacteria may result in a drastic breakdown of glue even though holding temperatures are maintained at higher levels. *Preservatives* may be used; but in such dilute liquors any small amount added at an effective level is multiplied many times when the glue is finally concentrated. Formaldehyde has been widely used because of its low cost and effectiveness, but its use must be carefully controlled because of its slow but definite tanning effect on the glue protein. This action is manifested first by an increase in viscosity. Larger amounts of formaldehyde then produce a crosslinking which eventually leads to gelation and insolubility. Low levels of formaldehyde, as well as chromium and aluminum salts, have been used to obtain controlled increases in glue viscosities. Such viscosities are "false consistencies," but are sometimes used to gain water-taking capacity for increased coverage or spread. Hydrogen peroxide is a most effective preservative in liquors of low pH and was used for this purpose in gelatin processing for many years until faster schedules and air-conditioned dryers made its use unnecessary. Its greater importance has been as a bleach.

Common *bleaching agents* are sulfur dioxide and sodium hydrosulfite (Lykopon). Gaseous sulfur dioxide has been used to clarify and bleach animal glue for sizing material. Light liquors are gassed with sulfur dioxide until a pH of 4.7–4.9 is achieved. At this point many impurities, including suspended colloidal material, fat, and mucins, coalesce as a floc and settle. The very light colored supernatant clear liquor is then evaporated to remove excess sulfur dioxide. There is relatively extensive hydrolysis during this treatment, which restricts its use to last-run liquors of low jelly strength. Each step of light-liquor processing has to be handled promptly to avoid bacterial and enzymic breakdown. Even judicious use of preservatives cannot overcome the disadvantages of excessive holding times or processing delays.

Concentration of light liquors takes place in multiple-effect vacuum evaporators of the Swenson or Yaryan types (see Evaporation). Such commercial film-type evaporators can concentrate only to a given limit determined by the viscosity of the concentrated or heavy liquor. If a given liquor is concentrated beyond this limit of viscosity there will be a hydrolytic breakdown so that higher solids are obtained at the expense of viscosity or quality. Early runs containing the higher-testing glue are usually concentrated to 8–18% solids at which point they gel readily upon cooling. Later runs of lower test may be concentrated up to 35–50% solids. More recently a new type of concentrator has been developed, which removes the evaporating film by a film-thickness regulator (1). Such evaporators operate on the agitated-film principle and are particularly efficient in the concentration of viscous liquids without destruction of the viscosity. They make possible the concentration of high-quality glue liquors to a solid content of 40% with maximum economy in the final drying step. Heavy liquors are much less susceptible to spoilage and after the addition of 1–1.5% zinc sulfate, based on glue weight, the heavy liquors rarely deteriorate throughout the remainder of the processing. Other additives are usually introduced at the heavy liquor stage. Defoaming compounds, which are normally fatty emulsions (sulfated tallow, etc), are added to the extent of 1–2% based on dry glue. They control foam when the glue is to be used on high-speed machines.

Drying of glue can be carried out in several ways, all of which are employed commercially.

Wind-Tunnel-Dried Glue. Heavy liquors, concentrated to the point of rapid gelation when cooled, are spread on a rubber belt which passes through a refrigeration chamber. A continuous strip of glue jelly is formed. This is then cut and spread on wire frames or nets which are subjected to streams of cool air until the gel has "skinned over." The air temperature is then raised and the process continued until the sheet of glue jelly is dry, hard, and brittle. Drying by the wind-tunnel method requires from two to four days depending upon the humidity, and the size of the gel pieces or ribbons. Continuous drying may be applicable when extremely thin films are produced. Conditioned air simplifies this drying operation since excessively fast drying at very low humidities causes fractional precipitation of sulfates and phosphates in regular layers (Liesegang rings), creating an effect known as "sandwich glue."

Scheidemandel (Smooth) Pearl Glue. In this procedure, developed in Germany, the concentrated heavy liquor is extruded through "droppers" so as to form beads. These are dropped into the top of a tower containing chilled kerosene or light oil. As they fall they form small "pearls" of jelly which are dried in a fluidized bed first with cold and finally warm air. The oil on the surface of the pearls evaporates during the drying operation. The final product is a smooth beadlike product free of fines and foreign particles.

Coated Pearl Glue. Evaporated heavy liquors are dropped or punched into small pearls which fall upon a moving bed of granulated dry glue, and are then covered with another layer. This sequence is repeated until the belt carries a bed about 3–4 in. thick, each wet pearl being thoroughly coated with dry granules of the same type. The product is then screened, the granulated bed glue returned for recirculation, and the soft coated pearls proceed to the hot-air dryer. Normally the amount of dry granulated glue picked up by the wet droplet will be approximately twice the weight of glue contained in the drop. More finely ground bed glue will result in a lower level of "pickup" (2).

The development of concentrators, as previously described, has made possible the production of very highly concentrated liquors without loss of quality. Such liquors can be gelled almost instantly by moderate cooling. This makes possible the use of chill rolls from which the jelly can be cut into fine pieces and dried rapidly by tumbling in air.

Irrespective of the drying procedure employed, the final product is ground to size and blended to a given test and grade.

Chrome glue is a form of hide glue obtained from chrome leather scrap with some minor changes in the extraction procedure. Pieces and shavings from chrome-tanned leather, having been thoroughly lime cured prior to tanning, require only the removal, or else the fixation, of the chromium before extraction of the glue. The chromium salts are removed by treatment—consecutively and preferably countercurrently—with caustic soda and hydrochloric acid. The swollen dechromed stock is then ready for extraction. Fixation of the chromium is also effected by conversion to an insoluble alkaline earth chromite. This may be done by prolonged liming, or even better by magnesium oxide or magnesite. In either case the excess alkaline earth is washed out and the stock extracted in the usual way. Chrome glue is particularly suitable for uses in which high jelly strength and clarity are desired. It contains comparatively little fatty material and may be formulated into very clear liquid and flexible adhesive compositions.

Technical gelatin (or inedible gelatin) requires increased clarity and lightness of color, and for this reason both light and heavy liquors are filtered using filter aids, as well as cellulose pulp, for selective adsorption of traces of fat. Bleaching is usually effected by peroxide rather than sulfur dioxide or Lykopon since there is less deterioration. Traces of nitrate ion greatly enhance the bleaching action of the peroxide. Since technical gelatin is necessarily relatively high testing, it is rarely concentrated beyond 6 or 8% solids prior to drying. This results in a very thin clear dry sheet and accounts for the expression "thin cut" as a synonym for technical gelatin.

Bone Glues

Green Bone Glue. The raw materials are supplied fresh from the meat processing plants. The quality of bone glue varies greatly with the kind of bone used and each type is usually handled separately. Killing bones (soft bones) are those removed on the dressing floor and consist of skulls, jaws, and feet. Cutting bones (hard bones) are those trimmed out of carcass meat after chilling and include ribs, plates, chucks, knuckles, loins, butts, rump, and pelvic bones. Cutting bones yield a better quality glue, probably because of the prompt handling and refrigeration they receive. Cattle feet are the exception since they are handled in a preliminary rendering operation to recover neatsfoot oil. The residual hard shin bone yields a very excellent glue product.

At the glue factory the bones are simultaneously washed and crushed so as to remove dirt, blood, fat, and extraneous matter. Then they are charged into pressure tanks of 12,000–30,000 lb capacity containing boiling water which serves to preheat, shrink, and curl nonbone collagenous material (some sinews, cartilage, and skin). This change in physical form (curling) prevents packing and channeling during subsequent extraction procedures. The preliminary hot wash extracts no glue and the water is then run off and handled for recovery of grease only. The pressure tanks are closed tight and direct steam is introduced under pressure so as to cure the collagenous constituents by steam rather than acid or alkali. The amount of pressure and the length of time it is applied vary widely among manufacturers and are dependent upon the type of glue desired. After the pressure cook and withdrawal of condensate the cooked bone material is covered with boiling water for about 1 hr before drawing off the first run. A second and third run may be taken as well if the extracted solids remain sufficiently high. A second pressure cook is then made and another series of runs extracted. This is repeated until all possible glue has been extracted. The small volume of residual bone remaining in the tank is finally just covered with hot water and given a longer cook at considerable pressure without removal of water. This is called a "stick" cook because it completely hydrolyzes all the remaining protein and leaches it from the residual "steam bone meal" as a hydrolyzed proteinaceous liquor, without any appreciable jelly strength. The name is descriptive of the material after evaporation.

Owing to the hygroscopic nature of protein hydrolyzate, "stick" is never dried alone but mixed with tankage, blood, or other protein material to produce a dry granular product suitable for animal feed formulations. Such products may contain as much as 65–70% protein. In order to reduce this to 50–60% as desired, steam bone meal is blended in. When sold separately steam bone meal is valued according to its BPL (bone phosphate of lime or tricalcium phosphate) content.

Just as with hide glue, the early runs of bone glue are of the best quality. They are processed in the same way as hide liquors, namely, drawn off, settled, skimmed, evaporated, and dried. Bone glue liquors used for colloidal rather than adhesive purposes are frequently bleached, centrifuged, clarified, and filtered for maximum removal of fat, color bodies, and other nonglue solids.

Extracted Bone Glue or Dry Bone Glue. Any bone material may be converted into a suitable form for this type of process by cooking and drying or, in arid climates, by drying alone. Much of the livestock in South America, for example, is processed into frozen meat for export. There is a large tonnage of bone by-product, which may be handled by any one of the following methods.

Drying of the fresh bone by spreading results in slow dehydration and development of oxidative rancidity in the fat. Finally sufficient breakdown occurs to produce aldehydes and render the bone protein insoluble. For this reason degreasing is necessary for a satisfactory glue raw material. A short cook in boiling water is usually sufficient to render and float the major portion of fat without appreciable loss of glue protein. Open-air drying then produces very desirable raw material. Acidic water aids in this degreasing operation and the final product is called "acidulated" dry bones. When dry bones are stored for some time before shipping they should be degreased with fat solvents. Thus raw material is obtained superior in yield and final quality of glue. Hence the name "extracted bone glue" is applied to the excellent product obtained. Acidulated or extracted bones give the best yields of high-test bone glue and for many industrial purposes this can be blended with the more expensive hide glues for economy.

Ossein Glue. Strictly speaking, this is another form of bone glue but it is normally treated as a raw material for gelatin. When bones are treated with dilute hydrochloric acid the insoluble mineral portion, consisting essentially of tricalcium phosphate, is converted into the soluble calcium acid phosphates and subsequently washed out. This process leaves a residue of the bone connective tissue collagen, referred to as "ossein," which lends itself to conversion into either glue or gelatin. using the procedures which have been described for hide glue.

Fish Glues

Two types of fish glues are made from by-products obtained from the processing of fish for the fresh and frozen seafood markets. The highest-quality product is called *isinglass* or *fish gelatin* and comes almost entirely from the swimming bladders or sounds. The second type, called *fish glue*, is a much less expensive product and is extracted from the heads, skins, and skeletal waste from cod, haddock, mackerel, etc. Glues of the second type differ from animal glue in that they have no great tendency to gel and are readily extracted by hot water without preliminary acid or alkali curing. The raw material is thoroughly washed to remove all salt, dirt, and blood and then extracted by cooking in hot water in open kettles. Pressure cooking may be used as well. The dilute liquors are allowed to settle to facilitate the skimming of grease and, after bleaching with peroxide or sulfur dioxide, are filtered and concentrated to viscous liquids in which form they are sold.

Isinglass is not extracted in the same way as fish glue although it may be converted into fish glue by hydrolysis. Instead the fish swimming bladders or sounds are opened, washed, stretched, and dried to produce the isinglass of commerce. Fine sheets of this protein are produced by pressure rollers after softening in cold water. These may be shredded or flaked for use as adsorbents for impurities in the clarification of wines and other beverages.

Fish glue, which for many years was the only cold liquid glue available, has been largely replaced by many improved adhesive products derived from numerous synthetic resin raw materials. Today by-products from which fish glue was extracted are now processed into fish meal and fish solubles, both of which are in great demand as animal feed materials.

Yields of Animal Glues and By-products

Residues remaining after the extraction of glue from hide materials are fat and grout. The fat is skimmed off wherever possible to retain its maximum quality. This does not apply to the fat which is held by the grout. Grout is the nonsoluble residue remaining in the extraction kettle after all hot-water-soluble glue has been dissolved. It consists of tissue proteins, elastin, meat fibers, and hair along with a large

Table 1. Yields of Glue and By-products from Various Hide Raw Materials

	Yield, %		
Raw material	Glue	Fat	Grout
calf stock	14–16	3–4	5–7
bacon skins	24–26	26–34	2.5–4
cattle tails	14–15	3–4	15–16
chrome stock	25	none	none[a]
coney	45–60	none	none
dried hide trimmings	30–35	none	10–12
green fleshings	6–8	15–20	4–5
hide trimmings	15–16	3–5	7–9
sheep stock	6–7	6–9	3–4
green snouts, lips, ears	12–14	5–7	12–14
sinews	17–19	6–9	5–7

[a] Discarded.

quantity of fat not previously skimmed off, much of which has been fixed as lime soaps. It is processed by heating with dilute acid to decompose lime soaps and liberate fatty acids, skimming off the crude fatty acid material, and processing the remainder with other tankage materials. From bone glue, only fat and steam bone meal are obtained. Table 1 illustrates the average yields of glue and by-products from various hide raw materials.

Salted raw materials with a lower moisture content yield proportionately more glue. Wide ranges result because of variation in the condition of these perishable raw materials.

The bone raw materials are either green bone fresh from the meat-processing unit or dry bone which has been previously cooked or extracted to remove fat. Green bone yields about 12–13% glue, 12–15% fat, and 30–35% bone meal. Dry bones yield much more glue (18–20%), almost no fat, and a large quantity of low protein bone meal (65–70%).

Properties of Animal Glue

Animal glue—as also gelatin—is a hydrolysis product of collagen and comprises many fragments which vary in size from simple polypeptides to rather large colloidal molecules. This range corresponds to variation in molecular weights from 3000 to 80,000 approximately. The mixture can be fractionated by partial precipitation with methanol, ethanol, or acetone to yield molecular aggregates somewhat more uniform in size.

The molecules of larger size are believed to be spirally round helical chains connected through hydrogen bonding. Whether the average molecular weight is 20,000 or 80,000 the number of free amino and carboxyl groups remains constant at approx 45 and 75 milliequivalents, respectively, per 100 g of the sample. These values do not change appreciably until a major hydrolytic breakdown has taken place with rupture of peptide linkages. Therefore such physical properties as viscosity and gel strength are a better indication of molecular size than the conventional Van Slyke method for determining free amino or carboxyl groups.

The color and odor of the product depends largely upon the care given the raw materials prior to delivery to the glue manufacturer and upon the processing methods used for extracting the glue proteins. Light color and clarity usually indicate high quality, with use of bleaching and filtering. The actual amount of breakdown is reflected by the jelly strength and viscosity of the product with due recognition of the fact that both these physical constants can be adjusted by chemical reagents. The horny, resinous external appearance of the dry ground product creates the illusion that the amorphous protein is crystalline. Animal glue does not have a melting point but when heated rapidly to 200°C it will expand into small spheres due to the rapid evolution of moisture. Slow heating leads to decomposition and finally charring. Heating of the by-product at 80–100°C for a number of hours causes gradual dehydration with condensation. This is shown by a gradual increase in viscosity until the product becomes insoluble after heating at 110°C for about 12 hr.

Boiling water hydrolyzes the protein slowly at the neutral point but this breakdown is greatly accelerated by acids or alkalies. The proteolytic enzymes papain, bromelain, ficin, and trypsin hydrolyze glue rapidly, with a corresponding increase in the amino and carboxyl groups liberated. Pepsin acts differently with the viscosity falling off very rapidly; but relatively little change in the number of free amino groups indicates that few peptide linkages are broken by this enzyme. Enzyme reactivity varies widely with previous hydrolytic degradation (3).

Animal glue is insoluble in cold water but absorbs from 6 to 8 times its weight of water at temperatures below its congealing point. For this reason solutions of animal glue are always prepared by soaking the material in cold water for a time followed by warming to 50–60°C at which point the gelled particles pass smoothly into solution. Water is the only true solvent for animal glue but there are many cosolvents which are effective in the presence of some water. These are glycerol, glycols, sorbitol, acetic acid, formic acid, formamide, dimethylformamide, dimethyl sulfoxide, methylbutynol, methylpentynol, and *N*-methylpyrrolidone.

Chemical reagents which normally react with amino, hydroxyl, mercapto, and carboxyl groups react with these groups in glue. Glue proteins may be *O*- and *N*-acylated by fatty acid anhydrides or acid chlorides. This reaction with higher fatty acid chlorides has been used to prepare highly surface-active materials not only from glue hydrolysates but other proteins as well.   Nitrous acid generated in situ effectively deaminates the glue with conversion of almost all terminal amino and amide groups to hydroxyl and carboxyl respectively.   Isocyanic acid similarly converts the amino groups to urea groups which then react with aldehydes quite differently from the parent protein (4).   Symmetrically disubstituted carbodiimides cause a secondary condensation of the protein by loss of water and formation of additional amide linkages (5).

Reactive aldehydes, such as formaldehyde, glyoxal, glutaraldehyde, and even the reducing sugars, set up a chain of methylolamine reaction products which continue to condense until final crosslinking has converted the solution into an irreversible gel. At this stage the product is insoluble but if complete drying has not been effected, this reaction can be reversed with an aldehyde scavenger such as urea or thiourea. These aldehyde reactants have been used to increase the viscosity of animal glue products, but this practice is not recommended because it is difficult to control this reaction.

Heavy-metal polyvalent ions normally used in leather tanning operations, such as chromium, iron, zirconium, and aluminum, react with animal glue, forming an insoluble product or an irreversible gel.   These reactions are usually regulated by controlling the pH.   In addition to the reactive aldehyde and polyvalent metal salts, other tanning agents, both natural and synthetic, render insoluble or "tan" animal glue.   This product differs from many other proteinaceous materials because it does not coagulate at a given temperature and is most difficult to precipitate even at its isoelectric point.   Glues from acid cured stocks have an isoelectric point of about 8–8.5 and from alkaline cured stocks of about 4.5–4.8.   Protein precipitants most effective for glue are trichloracetic and phosphotungstic acids.   Reactions which convert animal glue into an insoluble material seem to have no effect upon its moisture resistance.   For example, formaldehyde tanned glue swells in cold water quite readily but does not pass into solution when warmed.   This is explained by the small proportion of the large protein molecule which is involved in the crosslinking reaction.   On a mole for mole basis a glue with a molecular weight of 30,000 would be made insoluble by about 0.1% formaldehyde; therefore, moisture sensitivity remains after total insolubility is achieved.   The same can be said for almost all glue reaction products. Epichlorhydrin and diepoxybutane are bifunctional compounds which crosslink animal glue very rapidly.

Perhaps the most characteristic property of animal glue is its ability to form reversible gels in water solution.   The rigidity of these gels increases with lowering temperature and rising concentration.   The melting point, as well as rigidity of the gels, also depends upon the molecular size or quality of the glue.   Numerous agents exert a pronounced effect on the gelling tendency of animal glue.   Of the soluble salts, chlorides and nitrates retard gelling or prevent it.   In this respect alkaline earth salts are more effective than the alkali metal salts.   Sulfate and phosphates exert little or no effect. Zinc sulfate does not affect the jelly strength of a glue even at high levels, whereas sodium chloride will decrease the jelly strength and calcium nitrate will amost prevent gelation.   These salts are used frequently to control melting point, gelation, and setting time in adhesives based upon animal glue.   Citrates, tartrates, and malates tend to increase the rate of gelation but not the rigidity of the gel itself.   Many other liquefiers or peptizing agents are found among the organic compounds, such as urea, thiourea, sodium 2-naphthalenesulfonate, chloral hydrate, phenol, and many of the phenolsulfonic acids.   Ethylene chlorohydrin is an example of a volatile liquefier.   Its use is limited by its reaction with amino compounds and by the ease with which it may be hydrolyzed into the nonvolatile glycol.

Specifications and Standards. All dry animal glue is sold as hide glue, bone glue, or a mixture of both. Its increased use as a protective colloid has led to more descriptive terms, particularly when purchased for specific nonadhesive uses. Such expressions are technical protein colloid, protective colloid, chrome glue, dry bone glue, and green bone glue, depending either upon its field of usage or the raw material from which it has been derived. All animal glue has approximately the same percentage of moisture (8–10%), ash (2–6%), fat (1–6%), and glue protein (80–90%). Its general quality and hence utility, depends upon the amount of hydrolytic breakdown that may have taken place during its manufacture, and this is reflected in such physical properties as solution consistency and gel rigidity. Peter Cooper, one of the earliest of commercial glue manufacturers in the United States, recognized the need for a uniform method for measuring these qualities and in 1844 developed a system for grading animal glue. This was based upon the stiffness or rigidity of a gelatin gel of definite concentration and at a definite temperature. Several subsequent attempts were made to set up a procedure which would set a numerical value for the jelly strength of a glue. Such devices were introduced by Lipowitz (1861), Kissling (1892), and Valenta (1909). It was not until 1923 that Oscar Bloom developed the gelometer which bears his name. Shortly thereafter its success in giving reproducible results led to its adoption as the standard device for determining gel strength in grams (Bloom test) (see p. 506).

Many glue manufacturers affix descriptive trade names to their glue products and assign a range of jelly strength with which they must comply.

Today the numerical Bloom test in grams is universally used to the exclusion of almost all other specifications, except in a few very special applications. Individual users may choose to specify additional limitations on ash, fat, and foaming tendency since these may be important in the specific process or product with which they are concerned.

Test Methods. Standard procedures have been developed for the determination of the following constants or impurities:

Jelly strength and *viscosity* give an excellent picture of the grade of glue, and the ratio between these two properties serves to indicate the source material and the process by which the glue was extracted. These two values are determined on the same sample by the following procedure (6):

Air-dry glue (14.3 g) is soaked in 100 ml cold water (10°C) until thoroughly hydrated. The granular gel is melted in a water bath at 60°C and is run at this temperature through a standard pipet calibrated against water. The viscosity in millipoises is calculated from the seconds of flow. The solution is then returned to the standard glue jar to test for rigidity or gel strength by determining the weight in grams required to press a flat circular plunger (1 cm² in area) a distance of 4 mm at 10°C into the jelly. Reproducibility of 1% can be expected.

Normally the ratio of jelly strength in grams to the viscosity in millipoises is about 3:1 for alkali-cured hide glues (eg, 300 g and 100 mP). Extremely high-testing alkali-cured products may exhibit a 2.5:1 or even a 2:1 ratio (eg, 400 g and 160–200 mP). The same raw material processed through an acid cure has a somewhat higher jelly strength and a much lower viscosity so that the ratio will be 4 or 5:1 (eg, about 450 g and 100 mP). Green bone glues usually have a 2:1 ratio (eg, 150 g and 75 mP) and extracted or dry bone glue a 2.5 or 3.5:1 ratio (eg, about 225 g and 75 mP). In the evaluation of results it should be remembered that viscosity is easily increased artificially so that "false" conclusions may be drawn unless a more complete analysis is made.

Grease content is important when the product is to be used in paper coating, sizing, or laminating. An excessive quantity of fat or grease impedes spreading over a smooth surface producing "fish eyes" or small areas which remain uncoated. Solvent extraction of the dried glue gives low results because of entrapment within the protein. The Kissling method is preferred since it involves complete acid hydrolysis of the protein, thereby releasing all fat, and subsequent extraction of the hydrolysate with petroleum ether.

Foam or rather tendency to foam is measured by a number of methods all of which employ vigorous mechanical agitation to entrap air, and measurement of the time required for the foam to recede or disappear.

There are specific procedures for determining various adhesive properties of animal glue (5). These include the determination of the rate of set, rate of drying, shear, cleavage, and tensile strengths, etc.

Economic Aspects

Representative figures for animal glue production in the United States reflect a great deal about the general economy and the effects of substitute materials that have entered the market. Since fish glues are less than 5% of the total animal glue produc-

Table 2. Animal Glue Production in the United States, million lb

Year	Hide glue	Green bone glue	Extracted bone glue
1922	57.9	28.2	8.5
1929	54.3	37.2	14.8
1947	67.9	71.3	18.1
1956	57.0	48.5	10.8
1960	50.9	40.4	11.9
1961	48.6	35.6[a]	8.8[a]
1962	48.8	40.0[a]	9.9[a]
1963	50.7	32.7[a]	8.2[a]

[a] In 1961–1962 extracted bone and green bone glues were reported together so that these figures are calculated from totals on the assumption that 20% is of extracted-bone and 80% is of green-bone origin.

tion their figures have been grouped with other miscellany and are not available after 1938. Table 2 gives the figures for animal glue production in the United States, compiled from data published by the U.S. Tariff Commission and the Department of Commerce.

It is of interest to note that the gradual drop in animal glue production coincides fairly well with the tremendous growth of various resinous adhesive products since World War II. Production has not fallen off more severely because animal glues are used today more as protein colloids than as adhesives.

Uses for Animal Glue

Two properties of *hide glue* account for the bulk of its use, its gel strength and the extreme strength and toughness of its dried film. The stiffness and resilience of hide glue jellies, particularly in concentrated 80% glycerol solutions, have led to its use in printer rollers, padding adhesives, cork compositions (gaskets), and shock absorbing pads. If insolubilized with formaldehyde, these concentrated glue–glycerine gels are useful to neutralize machine vibration since they are virtually unaffected by oil and grease. The other important property of hide glue is its film strength, which makes it ideal for wood joining operations, adhesion of grit to abrasive paper and grinding wheels, and the sizing of textile fibers, paper or printed strings used for gift wrappings, flowers, etc. Hide glues specially processed for grease removal can be made to form stable foams which gel when cooled. This property is useful in the formulation of match head combinations, since the protein is instrumental in retarding afterglow when the flame is extinguished.

Bone glues are used generally as adhesives when high film tensile strength is not required. In certain areas, such as protective colloid uses, hide and bone glues may be used interchangeably. Otherwise the adhesive properties of bone glue render it particularly suitable for paper laminating and sizing, as well as gummed stock to be activated through remoistening (paper labels, gummed tape, etc). Higher-testing bone glues are equally suitable for many hide-glue uses and in recent years there has been less insistence upon one or the other, as long as the animal glue product performs satisfactorily and complies with the user's specifications.

Perhaps the greatest change in glue usage over the past twenty years has been the transition in its importance as an adhesive to that of a protein colloid. Control of coalescence of tacky materials, flocculation and recovery of suspended solids, flattening agents for dyeing or metal plating, etc, all make use of the colloidal rather than the adhesive properties of glue.

Animal Glue Specialties. *Liquid glues* are prepared from all types of animal glue by the depression of the temperature of gelation with peptizing agents in much the same way as the freezing point of water is lowered by the addition of salts. Such "liquefiers" include urea, thiourea, sodium naphthalenesulfonates, ethylenechlorohydrin, chloral hydrate, ammonium thiocyanate, and the chlorides or nitrates of calcium, magnesium, or zinc. Most liquid animal glues contain about 35% water, 45–55% glue, and 10–20% liquefying agent, depending upon the minimum temperature at which the liquid state is to be maintained. They may be produced conveniently by adding the reagents to the heavy liquors and concentrating under vacuum to the desired solids content. An alternate procedure is to mix all ingredients and agitate slowly at ambient temperatures or with moderate heat. Application of these liquid glues by machine tends to produce long strings or threads (cottoning or feathering), due primarily to the failure of the fluid to gel as it sets. High-speed machines may produce so much of this weblike material that other ingredients must be incorporated to "shorten" the product. Such materials are the various flours and starches (wheat, corn, sago, tapioca, etc) which not only eliminate the stringing but increase the tackiness, thereby improving the function of the glue in high-speed labeling adhesives. Liquid glues to which urea or thiourea have been added are tough rather than brittle because of their tendency to retain moisture. They may be added to ordinary hot glue solutions in order to reduce the rate of set. Their main advantage is their ability to remain fluid at room temperature. A major disadvantage is the loss of viscosity upon aging several months at room temperature with a corresponding drop in adhesive strength.

Opaque glue is a less important specialty manufactured for specific application where a pigmented glue line is desired (eg, picture frames). White pigments, such as the oxide of zinc or titanium, are milled with a small quantity of glycerine to form a smooth paste which is then dispersed into the heavy glue liquor to produce opacity before drying in the usual way. Other color tints may be used as well to produce any shade desired.

Flexible glues, as implied by the name, are concentrated glue jellies containing a humectant and are used mainly as hot-melt adhesives. A wide range of flexibility can be attained by controlling the amount and quality of glue, the amount of hygroscopic agent, and the amount of water. The rate at which the product sets after application is dependent upon the quality of the glue base, the temperature, and the rate at which it loses water in the presence of the hygroscopic agent. The type of animal glue used in flexible formulations may be hide or bone or both and, depending upon the use, may constitute from 10% to 70% of the formulation. High-temperature use requires a product free of sugar because of the tendency of the sugar to make the glue insoluble through a conventional aldehyde reaction. Therefore, manufacturers of printed rollers and magazine binderies are prone to use only glycerine or sorbitol in their compositions. Even glycols are avoided because of their tendency to bleed into the paper edges and promote embrittlement of the backing or cover. In addition to such conventional polyols as glycerol, glycol, and sorbitol, other humectants may be preferred in those adhesives used more for their nonwarping properties than for perpetual flexibility. Among these are the simple sugars derived from the hydrolysis of starch, cane or beet sugar, invert syrup, lactic acid salts, etc. Such formulations are reasonably fluid when melted, contain enough animal glue to maintain the product in cake form, and after melting can be diluted with water. These solutions are applied hot to paper or fabric to be subsequently laminated to cardboard or wood, and do not warp the coating material. For this reason many of the low-cost flexible glues are called "nonwarps" as sold to the paperbox industry.

Truly flexible formulations are needed for printer rollers, padding glues, shock-absorbing pads, and gaskets and crown closures for bottles; since high tensile strength is equally important these uses require high-test hide glue. Nonwarp uses, such as flocking, lamination, and paperbox covering, and in luggage, table pads, etc, are formulated from either bone or hide glue or both, and use the sugars to impart nonwarp characteristics during use but rigidity in the finished product.

Bibliography

"Glue" in *ECT* 1st ed., Vol. 7, pp. 207–215, by H. H. Young, Swift & Company.

1. A. B. Mutzenberg, R. Fischer, and N. Parker, *Chem. Eng.* **72** (19), 175–190 (1965).
2. U.S. Pat. 2,024,131 (June 12, 1935), R. C. Newton et al. (to Industrial Patents Corp.).
3. J. H. Northrup, *J. Gen. Physiol.* **4**, 57 (1921b).
4. U.S. Pat. 3,021,321 (Sept. 8, 1958), H. H. Young et al. (to Swift & Co.).
5. U.S. Pats. 2,938,892 (Oct. 10, 1958) and 3,098,693 (July 23, 1963),
   J. C. Sheehan (to Arthur D. Little, Inc.).
6. National Association of Glue Manufacturers, *Ind. Eng. Chem. Anal. Ed.* **2**, 348–351 (1930).

General References

Alexander, J., *Glue and Gelatin*, Chem. Catalog Co., New York, 1923.
Alexander, J., *Colloid Chemistry*, Chem. Catalog Co., New York, 1932.
Altpeter, J., *Die Patentliteratur der Eiweisstoffe*, Allgemeiner Industrie Verlag, Berlin, Germany, 1932.
Ames, W. M., *J. Soc. Leather Trades' Chemists* **33**, 407 (1949).
Animal Glue in Industry, National Association of Manufacturers, New York, 1951.

"Animal Glues: Their Manufacture, Testing, and Preparation," *U.S. Dept. Agr. Forest Service, Forest Products Lab. Rept. No. D492*, 1948.

"A Water Resistant Animal Glue," *U.S. Dept. Agr. Forest Service, Forest Products Lab. Rept., No. R40*, 1933.

Bogue, R. H., *Chemistry & Technology of Gelatin and Glue*, McGraw-Hill Book Co., Inc., New York, 1922.

British Standards Institution, *Brit. Std. No. 647* (1938).

Delmonte, J., *Technology of Adhesives*, Reinhold Publishing Corp., New York, 1947.

Gerngross, O., and E. Goebel, eds., *Chemie und Technologie der Leim- und Gelatin Fabrikation*, J. F. Steinkopf Verlag, Dresden, Germany, 1933; (photo-lithoprinted by Edwards Bros., Ann Arbor, Mich., 1944).

"Important Factors in Gluing with Animal Glue," *U.S. Dept. Agr. Forest Service, Forest Products Lab. Rept., No. R869*, 1929.

Kantrowitz, M. S., E. W. Spencer, and F. R. Blaylock, "Flexible Glues for Bookbinding," *U.S. Govt. Printing Office Tech. Bull., No. 24*, 1941.

Kissling, R., *Leim und Gelatin*, Wissenschaftliche Verlagsgesellschaft, Stuttgart, Germany, 1923.

Rideal, S., *Glue and Glue Testing*, D. Van Nostrand Co., New York (now Princeton, N.J.), 1914.

Sauer, E., *Leim und Gelatine*, J. F. Steinkopff Verlag, Dresden, Germany, 1927.

Simon, W., *Chem. & Met. Eng.* **84**, 101 (1938).

Smith, P. I., *Glue and Gelatin*, Pitman, London, 1929.

The Story of Animal Glue, National Association of Glue Manufacturers, New York, 1932.

Thiele, L., *Die Fabrikation von Leim und Gelatin*, Gebrüder Jänecke Verlag, Leipzig, Germany, 1922.

Tressler, D. K., *Marine Products of Commerce*, Chem. Catalog Co., New York, 1923.

HARLAND H. YOUNG
Swift & Company

APPLICATIONS

The number of uses for adhesives is very great, ranging from some industrial processes using tremendous total volumes each year to minor assembly operations that require small quantities, but nevertheless are very important from the standpoint of the end product. The use of adhesives entails certain advantages and disadvantages.

Adhesive bonding fulfills the need for an easy and rapid joining together of two adherends and in addition has several important advantages that are not found with mechanical fastenings, such as nails, screws, or bolts, with soldering, or with welding.

1. Certain materials that would be impossible or impractical to attach by other means can be joined. This advantage is utilized in the bonding of paper labels to cans or bottles.

2. Joining of two or more similar or dissimilar materials can be done more economically and efficiently by adhesive bonding than by other methods. The combining of many small pieces of wood into larger glued assemblies is an example of this.

3. Smoother surfaces and contours can be obtained by adhesive bonding of materials, such as brake and clutch facings or aircraft assemblies, in which rivet heads or other projections affect performance.

4. There is more efficient and uniform transfer of stresses from one member to another than with mechanical fastenings, as illustrated by modern glued-wood roof trusses, stressed-cover house panels, and helicopter rotor blades. Adhesive bonding has been particularly effective in increasing the fatigue life of assemblies by reduction of stress risers, as in rotor blades.

5. Adhesive bonding has made possible the design of new and better composite articles by taking advantage of the best properties of each of the joining materials. This is shown in the design of light but strong sandwich panels, flush doors, metal-faced plywood, adhesive-bonded rubber motor mounts, and endless combinations of wood, paper, metals, plastics, rubbers, and other materials.

6. Corrosion of joints in dissimilar metals is reduced by preventing direct metal-to-metal contact.

7. Better sealing action is provided for gases and liquids in joints than is possible with mechanical fastenings. Two examples of this are the bonding of extrusions to seal the edges of sandwich panels and the fabrication of watertight wood boats by adhesive bonding of components.

In addition, considerable reductions in weight are often possible using adhesive-bonded constructions, as in structural metal aircraft assemblies, by permitting use of thinner gages of metal that are reinforced where necessary, as around cutouts and along edges. Production costs can often be reduced by adhesive bonding as compared to conventional mechanical fastenings—for example, in certain aircraft assemblies. This is often particularly significant as the size of the assembly increases, since the entire assembly may be bonded at one time and is thus cheaper than automatic riveting. However, cost comparisons of adhesive bonding with other methods of fastenings are based on many interrelated factors, including the cost of the adhesive, related processing and equipment costs, labor costs, and other factors that make it impossible to generalize on actual relative costs of the different methods.

At the same time, certain limitations of adhesive bonding, as compared to the use of mechanical fastenings, must be recognized. Generally, relative strengths with adhesive bonding are somewhat more directional than with mechanical fastening, particularly with metal-bonding adhesives. Such adhesives are very good in shear, acceptable in tension, and low in resistance to cleavage or peel. Surfaces to be bonded must be cleaner and more carefully prepared than is usually necessary for mechanical fastenings. Adhesion of a certain chemical type of adhesive is generally better for some adherends than for others. Any given adhesive application, therefore, requires a rather specific adhesive formulation and may also require special bonding conditions. At present, there is no single adhesive or bonding process that begins to meet the requirements of a truly general-purpose adhesive. It is doubtful whether such a product is likely to be developed very soon. Adhesives generally do not develop their full strength and permanence immediately, as with welded or mechanical joints, but often require appreciable time for the reactions of setting or curing to take place. Production equipment is somewhat different from that for mechanical fastening techniques. Adhesive bonding requires jigs and presses to insure adequate and intimate contact of the mating adherends at the time the adhesive solidifies. Presently available adhesives, which are nearly all based on organic compounds, vary considerably in permanence in joints. Some types deteriorate as a function of time under certain conditions of service. For example, some have limited heat resistance above 400°F for periods of more than a few hours. However, recent developments indicate that much more thermally stable adhesives can be produced. Finally, the present lack of reliable nondestructive test procedures for evaluating the final bond quality in production makes it necessary to practice close quality control over the entire bonding process in order to be certain of uniform bond quality, particularly in structural joints.

Recent developments in new synthetic resins and other components have made possible the development of many new, stronger, more durable, and more versatile adhesives for bonding surfaces previously difficult or impossible to bond. In addition, a great deal of experience has been gained in improved bonding techniques and new bonding equipment has been developed. Greater attention has been directed toward better designs for bonded joints. This has resulted in wider, more important, and more critical applications than were ever thought to be possible. There is good reason to believe that this development of better adhesives and bonding processes will continue to an accelerated pace in the future.

The following discussion highlights a few selected applications of bonding.

Paper Adhesives. Paper bonding and related packaging applications represent one of the largest uses of adhesives at the present time. Adhesive bonding of envelopes, bags, cartons, gummed paper labels, and stamps is familiar to all. Earlier adhesives in these fields were formulated from hydrolyzed starch (dextrin), gum arabic or gum tragacanth, and certain modified casein or animal protein bases. Most of these were dispersed in water, and they developed strength by drying. Large quantities of sodium silicate and starch adhesives are used in bonding the paper liners to core components of corrugated fiberboards and laminated paperboards for packaging. Although these adhesives gave adequate strength for many applications, they often lacked the resistance to moisture and water required for overseas packaging. To improve these shortcomings, newer adhesives have been developed based on poly(vinyl alcohol) or poly(vinyl acetate), starch fortified with poly(vinyl acetate), and resorcinol or urea resins. Resorcinol-modified starch adhesives are a recent development for corrugated fiberboards with improved wet strength.

Hot-melt adhesives, based on poly(vinyl esters) and poly(vinyl acetals) or microcrystalline waxes, are finding broad applications in the paper field in the fabrication of milk cartons, heat-sealed labels, and in bookbinding. In bookbinding, the hot-melt adhesives have greatly speeded up the binding process and have eliminated the long drying periods that animal glues required for removal of water after binding. These drying periods slowed up the trimming of the pages and completion of the books. Before repulping the excess paper trimmed off the book pages after bonding, the hot-melt adhesives must be completely removed so that the properties of the new pulp are not adversely affected.

Pressure-sensitive tapes with cloth, paper, and transparent-film backings and various rubber and resin adhesive systems find wide applications in the sealing of cartons and shipping cases of paper and fiberboard. Recent high-strength tapes with glass-fiber or nylon threads in the backing provide strong bindings that can be used in place of steel strapping around heavy corrugated-fiberboard shipping cases. These tapes are superior to steel straps because they not only provide the necessary tensile strength and are easier to apply but they also prevent wearing and crushing of the fiberboard. They do this by distributing stresses over the entire bond area rather than concentrating them at corners and edges, as do steel straps.

Rapid-setting, carton-sealing adhesives based on sodium silicate or starch continue to find wide application in packaging. New paper products are now made possible by using adhesives of various resin and rubber bases to combine metal foils, polyethylene, and other plastic films with high-strength or wet-strength paper to make superior case liners, multiwall shipping bags, vapor barriers for building construction, and other items.

Wood Adhesives. Historically, the bonding of wood was one of the first uses for adhesives. Accounts of veneered wood and other glued-wood assemblies date back to the Egyptian pharaohs. Wood bonding is a very important means of improving the utilization of our wood resources. It is one of the largest-volume uses of adhesives, particularly in the softwood plywood industry of the Pacific Northwest, where an estimated 185 million pounds of dry adhesives was used in 1959. The older adhesives for wood were based on materials of natural origin beginning with the hide glues from animal protein by-products and then vegetable starch, casein, soybean, and blood adhesives. Although each of these types is still used in fairly large quantities, the newer synthetic-resin adhesives have been replacing many of these natural-base adhesives, particularly in certain applications. The principal uses for animal adhesives (hide glues) are in edge gluing of lumber for furniture and in the assembly of furniture. Polyvinyl resin emulsion adhesives have made important advances in these areas. Casein adhesives are used mainly for laminating heavy structural timbers for interior use in churches, schools, field houses, and commercial buildings. Soybean and blood adhesives are used primarily for bonding interior-type softwood plywood. Starch adhesives are still used to a limited extent in hardwood plywood for interior use but have been largely replaced by urea–resin adhesives that are often

Fig. 1. Glued laminated members achieve size, strength, and form with high-strength adhesives impossible with natural wood. Courtesy Weyerhaeuser Company.

extended with wheat flour for greater economy. Phenolic resins are used primarily for exterior-type softwood plywood, whereas resorcinol resins are used mainly for structural laminates for severe service, such as bridge timbers, crossarms, and exposed structures and boats (Fig. 1), and in fabrication of glued-wood components, such as trusses and prefabricated wall, roof, and floor panels for buildings. Melamine resins are primarily used to improve the durability of urea–resin adhesives in hardwood plywood for exterior service as in boats, a rather limited use at present.

There are a number of important new uses for adhesives, particularly in the lumber industry, that are under development or in early production trials and that promise to increase the use of adhesives for wood very significantly in the next few years. These involve either the combining of many smaller pieces of good-quality wood into standard sizes of panels for greater economy of use through edge and end gluing or the overlaying of lumber or plywood with special papers, hardboards, and metal or plastic sheets and foils to improve surface properties. Another growing use of adhesives is in the bonding of sheets of plywood, hardboard, or fiberboards to wood frames to produce high-strength, durable, and economical unitized panels for building components. Although each of these new products can be bonded satisfactorily with conventional wood adhesives, most of these adhesives are not readily adaptable to the economical large-volume production anticipated. There is great interest in the potentials of some of the new types of adhesives being developed for bonding other materials such as metals and plastics. Of particular current interest are certain of the rubber-base contact-setting adhesives originally introduced for bonding plastics and metals to plywood and other wood-base cores for interior use.

Metal-Bonding Adhesives. Largely because of the great importance of adhesive-bonded constructions in the aircraft and missile fields and the critical value of these fields in the cold war period, some of the most interesting and advanced technology in the entire adhesives industry has developed in structural metal bonding. There is every reason to believe that these efforts will continue to expand, and to expect that much of the basic information in this area will greatly stimulate other developments in the adhesives field generally. Modern aircraft structures, as in supersonic planes and missiles, place great emphasis on advantages to be gained from adhesive bonding techniques—including the need for smooth aerodynamic surfaces free of projections, the very efficient distribution of stresses from one member to another, the reduction in stress concentrations common to conventional mechanical fastenings, and the reduction in weight by stiff, rigid sandwich panels of thin metal skins bonded to lightweight cellular cores of metal or reinforced plastic honeycomb cores (Fig. 2). Extreme performance requirements could often be met best by adhesive-bonded components and thus spurred on the development of a number of novel adhesive systems of new chemical nature, of which the epoxy-resin adhesives are the most familiar at present.

Before World War II some metals, primarily aluminum and cold-rolled steel, had been bonded to wood with protein–rubber latex adhesives in a water dispersion. More recently the synthetic rubber-base, contact-bonding adhesives have been used for this purpose. These were essentially semistructural applications primarily for interior applications or for protected exterior uses. During World War II considerable development of new high-strength metal-bonding adhesives took place for limited applications in aircraft uses. At first these involved bonding aluminum to wood but later bonding aluminum to itself. The earlier resin adhesives were essentially modifications of the conventional phenolic resin adhesives used for wood by incorporation of neoprene or nitrile rubber or with certain thermoplastic resins, primarily poly(vinyl butyral) or poly(vinyl formal). Such modification was necessary to reduce the excessive rigidity and brittleness of the adhesive layer and to accommodate internal stresses in the joints caused by differential dimensional changes in the metals and wood. These dimensional differences are the result of temperature and moisture changes during service. The earlier adhesives, based on phenolic resin modifications, required curing at 300°F or higher. After World War II the early epoxy-resin adhesives were developed, some of which can now be cured at normal shop temperatures.

The earlier metal-bonding adhesives, based on combinations of phenolic resins and rubbers or thermoplastic resins, were used in solution form and problems were encountered in eliminating from the resultant joint residual solvent bubbles that reduced uniformity and joint strength. This has fostered the further development of film-type adhesives of the same general composition that require only insertion between the adherends and then hot pressing.

As service conditions for aircraft became more severe, particularly the temperature requirements, new metal-bonding adhesive systems were developed. The phenol–nitrile and phenol–epoxy combinations have generally been most promising for use at temperatures up to 500°F for periods up to about 200 hr. Such high-temperature service in aircraft, involving primarily new corrosion-resistant steel alloys rather than aluminum, introduces problems of thermal softening as well as thermal degradation of the component polymers. The B-58 (Hustler) supersonic bomber of the Air Force utilizes 4500 ft² of bonded panels and requires approximately 900 lb of

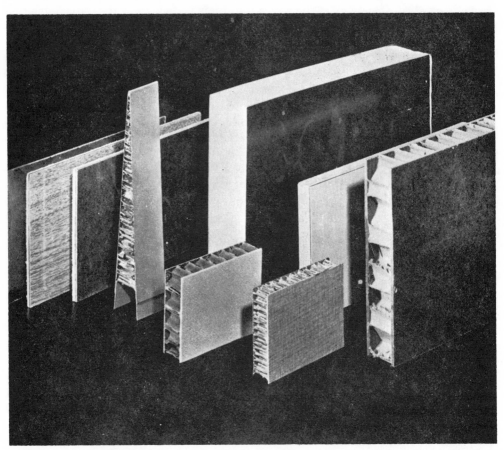

Fig. 2. Typical adhesive-bonded structural panels include plastic laminates to aluminum honeycomb, fiber glass to phenolic, aluminum to aluminum honeycomb, and asbestos–cement to plastic foam to aluminum. Courtesy National Starch & Chemical Corporation.

adhesive per plane. Such a construction is expected to withstand 30,000 hr of service at 325°F. Some of the newer planes have significantly higher temperature requirements. This introduces severe requirements for adhesive performance that will be difficult to meet. Some organic adhesives based on phenolic and silicone–phenolic systems are said to be promising for use at 1000°F for short periods. Work is in progress on semiorganic adhesives obtained by reaction of newer heat-stable organic polymers with inorganic reagents such as arsenic compounds. Some authorities consider that structural adhesives for performance for 1000 hr at 500°F in steel joints will be feasible soon. Other interesting work is in progress to develop inorganic and ceramic-type adhesives, based on modifications of ceramic frits, for even higher temperature performance.

Along more conventional lines, some of the present metal-bonding adhesives are being used to bond porcelainized steel to hardboard and other backings and to a variety of insulating cores for exterior building panels.

The automobile industry is beginning to use large amounts of structural adhesives in body fabrication to replace spot welding for much the same reasons that the aircraft industry uses adhesives. A current use is in bonding inner and outer shells of metal stampings for hood and trunk lids. The assembly operation is entirely automated and consists of the application of 80–100 spots of a special viscous adhesive by controlled ejector valves to the metal in its original oily condition. The panel assembly is mechanically crimped around the edges and a few spot welds are made to hold the panel together temporarily. Panels are subsequently processed through metal-cleaning systems, primer dips, and bake-oven cycles and finally the finish color coat and drying-oven cycles. The adhesive is cured to its final state during the baking of the primer. Advantages of the new bonding process include elimination of sound-deadener pads and spot weld patterns on the final surface of the finished panel. Corrosion of the panel is also reduced by the sealing action of the adhesive system.

A pump manufacturer claims to have reduced rejects very significantly by adhesive bonding of three separate metal die castings to form one complicated part. This replaces a one-piece casting that, because of its intricate shape, resulted in excessive blowholes in the castings. Tubular joints of metal are now adhesive-bonded for scaffolding, bicycle frames, and lines for refrigerants, gas, and hydraulic fluids. Epoxy adhesives are used to bond carbide tips to metal for cutting tools. The pressure bulkhead of a liquid ammonia tank in a research rocket aircraft has been bonded and sealed with a special metal-bonding film adhesive. A special epoxy–resin adhesive has been used to assemble water-tight, lightweight metal containers for shipping, housing, and storing complex electronic and mechanical gear for the armed services.

Brake and Clutch Bonding Adhesives. Two of the major automobile manufacturers have been bonding brake linings to drums since 1947. The principal adhesive system is apparently a phenol–nitrile rubber type used either in liquid or film form and heat-cured. Clutch facings and facings for automatic transmissions, where parts are immersed continually in warm oil, have also been bonded in recent years.

Printed Circuits. One of the principal processes for manufacture of printed circuits for electronic components involves the bonding of a copper foil to a phenolic-laminate panel using primarily vinyl–phenolic adhesives. Epoxy and nitrile–phenolic adhesives are also used. The circuit pattern is then applied to the copper film as a protective coating and the excess copper etched away. Various electronic components are later inserted through holes in the remaining metal and all soldered to the copper foil by a short dip in molten solder. The adhesive bond thus must retain its strength through the chemical etch and the hot solder dip, as well as in long-time actual service. The adhesive must have and retain electrical properties equivalent to those of the laminate base. It must not corrode the metal and should have no harmful residual solvents.

Bonding Ceramic Materials.   The epoxy-resin adhesives that set at room temperature have found a number of interesting uses in the bonding and repairing of various ceramic materials. These adhesives are used as tie coats between old and new concrete in repairing highways, bridges, dams, and similar outdoor constructions. They are also used for bonding traffic markers to highways. Epoxy-resin adhesives have been found to be very useful in gluing pieces of decorative marble, which often cracks when thin slabs are sawed from blocks, without destroying the patterns. Similar epoxy adhesives are used for bonding marble and other stone windowsills in place and have recently been proposed for bonding concrete blocks in basement walls.

Tile Adhesives.   Ceramic tile are now bonded to walls in large volume by using organic adhesives in place of the older inorganic mortar types. Adhesives include natural or reclaimed rubber bases in inorganic solvents, certain resin systems of unidentified chemical nature in solvents, and polymers or rubbers dispersed in water as emulsions or latexes.

Asphalt-base mastic adhesives, either as clay- or soap-stabilized emulsions or as cutbacks in hydrocarbon solvents, are widely used for bonding asphalt and vinyl–asbestos floor tile to wood or concrete subfloors in rapid on-the-job installations. Similar mastic adhesives are used for bonding wood floor tile to subfloors. The cutback types are used particularly where some water resistance may be needed, as on below-grade concrete slabs. Rubber tile and linoleum are bonded to subfloors either with low-moisture-resistant adhesives based on sulfite waste-liquor concentrates or with more water-resistant types based on certain natural resins dispersed in alcohol. Very viscous mastics of rubber or asphalt bases are used for bonding acoustical tile to ceilings.

Adhesives for Shoe Manufacturing.   A variety of quick-setting adhesives based on solvent dispersions of either rubber or cellulose nitrate are widely used in shoe manufacturing, particularly for women's shoes. In some applications the bond is intended as temporary until replaced by subsequent stitching or nailing. In other applicaions,, as in some ladies' shoes and sport shoes, the adhesive is the final fastening. Rubber-base adhesives are also widely used for bonding components in rubber footwear and for shoe repairing, particularly for attaching new soles and heels.

Assembly Adhesives.   In addition to a number of previously cited examples, adhesives of various compositions are used for attaching and assembling a wide variety of items. Many of these are used in very small quantities per item, but total volumes in some cases may be quite sizable. Some of these uses include attachment of rubber-gasket linings around automobile doors and trunk lids and installation of sound-absorbing pads in automobile bodies. Similar adhesives are used in attaching various upholstery fabrics and trim in automobiles. These are large-volume uses and require low-cost adhesives, such as those based on reclaimed rubber in hydrocarbon solvents. These adhesives must develop a minimum level of holding power almost instantaneously as the bodies move down a production line. On the other hand, epoxy-type adhesives and even the high-cost methyl cyanoacrylate adhesives have been used in small quantities for assembly of jewelry items and special optical and electronic components.   The growing interest in assembly of plastic model kits for hobbies has developed a sizable market for small-package, fast-setting adhesives.   Solvent-type adhesives based on polystyrene or cellulose nitrate are most important in these operations. Industrial modifications of these adhesives are also used in factory fabrication of plastic toys and household items.

R. F. Blomquist
Forest Products Laboratory
U.S. Department of Agriculture

EVALUATION

Because of the great variety of adhesives, adhesive applications, and bonded joints, the methods for evaluating adhesives are complex. Tests for structural adhesives are obviously more important and tend to be more elaborate and extensive than tests for holding or sealing adhesives. Generally, somewhat different methods of testing may be required for these various adhesive applications, but many methods have certain similarities. The potential industrial user of a new adhesive in any critical bonding operation will usually be concerned with the potential strength that can be developed on the adherends in the proposed joints; the working or operating characteristics of the adhesive while in use; the stability in storage before use; and the permanence of the joints produced. Before he decides to use the adhesive, he will probably want a fairly extensive study of its properties in order to be certain that the adhesive and the bonding process are capable of meeting the important requirements. These studies are often known as qualification tests. It is often important to make less extensive check tests on different lots of adhesives to be certain that each meets a certain minimum performance. These are often referred to as quality-control or inspection tests. Tests of actual production items may also be made to check the uniformity of the bonding process with a given adhesive over a period of time. Probably the most complete listing of test methods of adhesives in the United States is that published by the American Society for Testing and Materials (ASTM). These methods include procedures for various strength tests, for working properties, and for permanence. The ASTM compilation, which is revised at regular intervals, gives definitions of terms pertaining to adhesives as well as the test methods.

As previously noted, any strength tests of a bonded joint will depend both on the adhesive and on the bonding operation. With strength tests, particularly, it is therefore important to distinguish between tests of the adhesive itself as compared to tests of an adhesive-bonded product. In the first type the adhesive is used to fabricate a standard test specimen. Carefully selected and prepared adherends are used under controlled standard bonding conditions so that the resultant test values are a reliable indicator of the potentials of the adhesive itself. Tests on the bonded product from production are normally made on a standard type of test specimen. However, these are cut from an actual bonded item. Therefore, they may be of variable thickness and other variable dimensions and may be made up of different types or quality of adherends. They are then tested under the prescribed standard conditions. These test results are more a measure of the quality of the bonding operation, and often of the uniformity of the adherends, than an actual measure of the adhesive itself. Thus, in the case of adhesives for plywood, the adhesive is usually evaluated according to ASTM method D 906, whereas the plywood is evaluated according to ASTM method D 805, with both tests performed on the same type of shear-test specimen (1).

In strength tests, at least two distinct types of evaluation are of importance. First are the simpler test specimens and procedures that can be made rather rapidly and easily on readily available standard test machines and that give data which are primarily useful in the comparison of one adhesive with another under standard conditions. Such data may then be compared with similar data obtained at different times or even by different laboratories. An example of such a simple comparative test is the standard metal lap-shear test procedure described in ASTM D 1002 (1). It is readily recognized that these tests involve other types of stresses besides shear, and, thus, such data are not directly useful to design engineers. Therefore, a second type of strength test is needed to evaluate the adhesive in pure shear in order to satisfy the design engineer. Such a shear test, often proposed or used for this purpose, is a torsion test on metal tubes bonded end to end and loaded in torsion (2). Strains can be measured while under load, and stress–strain curves and various engineering calculations can be made. Such tests are essentially research studies that require much time and effort to make and test each specimen. The same distinction generally applies to tension, peel, fatigue, and other types of strength tests. Therefore, no single test procedure will satisfy both needs for information. Most of the published strength test procedures are intended for the simpler comparative tests. The more precise and fundamental test procedures have not yet been adequately developed or standardized and, therefore, are not usually included in present compilations of test procedures.

Initial strength tests of adhesives may include very simple knife, chisel, or pick tests by which bonds are crudely separated and the nature of the failure examined visually; more elaborate tension or shear tests in the case of rigid or semirigid adherends; or some sort of peel or pulldown tests where flexible adherends are concerned. Other more extensive tests may include impact, cleavage, torsion, dead load, creep, and fatigue tests, all of which give different but useful information.

Some actual test values in the simpler types of strength tests may be of interest for illustration. It should be noted that often these strength values are not adequate indications of the full potential strength of the adhesive system used. In the case of many paper-bonding and wood-bonding adhesives, formulations are usually developed to be as strong as the paper or wood. That is, the joint is expected to fail always in the wood or the paper, and this is the usual criterion of a good joint for these adhesives. Thus, the actual failing load will depend on the strength of the adherend and such adhesives are formulated to provide only this level of strength.

In one study, block shear test specimens were prepared with several different species of wood using one urea–formaldehyde resin adhesive that cures at room temperature (3). The specimen and test procedures were as prescribed in ASTM Method D 905-49. Table 1 gives the shear test values obtained.

Table 1. Shear Test Values

Species of wood	Shear strength, psi
Sitka spruce	1306
noble fir	1632
sweetgum	1769
yellow poplar	2143
black walnut	2393
yellow birch	2840
hard maple	3425

From the data one might infer that the adhesive was much stronger when used on yellow birch than on Sitka spruce. However, in all tests the specimens failed with estimated percentages of wood failure from 86 to 100, indicating that the strength of the wood was being tested rather than that of the glue. The shear strength of the particular adhesive was actually something greater than 3425 psi in these tests.

Various factors in the test procedure can influence test results. In the case of wood adhesives, these factors include the variability in wood pieces, moisture content, rate of load application, and variations in misalignment of specimens when under load. In one series of tests with the same block shear method (ASTM D 905-49) on hard maple, but using several early poly(vinyl acetate) emulsion adhesives, shear test values as high as 4156 psi were observed (4). However, over 90% of the failure was in cohesion of the adhesive film. This unusually high shear value on maple, without failing the wood, was attributed to some relief of stresses during loading because of the partial plasticity of the adhesive joint.

Structural metal-bonding adhesives, first developed extensively for use in military and naval aircraft, are usually evaluated by a simple lap-type shear test (ASTM D 1002-53T), using a specific aluminum alloy, 0.063 in. thick. Initial shear test values for a number of different types of such adhesives ranged from 1655 psi for a nitrile–phenolic adhesive to 4940 psi for a vinyl–phenolic adhesive. In most cases, failure was in cohesion of the adhesive film (5).

An extensive series of tests (6) of ten typical structural metal-bonding adhesives on the same aluminum-to-aluminum lap-shear test specimen (ASTM D 1002-53T) gave initial shear values at 72–75°F of 2496 psi for a neoprene–nylon–phenolic adhesive to 4561 psi for a vinyl–phenolic adhesive. Similar specimens tested at 178–182°F ranged from 934 psi for a neoprene–nylon–phenolic formulation to 4025 psi for a vinyl–phenolic adhesive. Similarly, when tested at −65 to −70°F, shear values were from 2414 psi for a neoprene-phenolic adhesive to 5090 psi for a neoprene–nylon–phenolic formulation. However, at higher temperatures, above 250°F, some of the other adhesives were much stronger than the vinyl–phenolic formulations.

In this same series of tests, similar aluminum lap specimens were subjected to axial fatigue at room temperature. The loads that could be tolerated without failure in 10 million cycles varied from 470 psi for a vinyl–phenolic adhesive to 1325 psi for a neoprene–phenolic adhesive. In some of the stronger specimens, the 0.063-in. aluminum alloy actually failed in fatigue in these tests.

Another important property of structural adhesives for aircraft is resistance to peeling. One of the most widely accepted tests for peel of aluminum-to-aluminum bonds is the climbing-drum test (7). In tests of 10 different structural adhesives with this procedure, average peel test values ranged from 6.3 in.-lb (1 in.-lb = 0.113 joule, absolute) for a liquid epoxy resin adhesive to 44.1 in.-lb for a nitrile rubber–phenolic adhesive.

These limited strength test values for a few typical wood and metal-bonding adhesives serve mainly to indicate the complexity of the test data already in the literature and that being developed further at a fast pace. The data are nearly all for the comparative types of tests rather than for the research or fundamental types of evaluation. The data indicate that formulating an adhesive for a given use is generally a matter of compromise among various performance requirements. Very seldom does the same adhesive in a group have the highest test values in each of several types of strength tests. These principles apply to adhesives for many different purposes.

There is often a tendency to stress strength properties of adhesives at the expense of other properties such as convenience of use, adaptability to a particular type of simple or rapid-bonding technique, low cost, or high permanence. The strongest adhesive is often not the most suitable for the job. For example, many mastic adhesives, formulated of rubber or asphalt, which are widely used for bonding resilient or wood floor tiles to concrete subfloors, may have shear strengths, measured by a modification of the block-shear test (ASTM D 905-49), of less than 50 psi, and yet they perform their assigned task entirely satisfactorily. Neoprene-based contact cements were developed largely for bonding high-density plastic laminates to wood-based counter or table tops. These adhesives develop a high degree of initial tack immediately after being assembled. They can then be rolled or tapped into place. The two adherends in this application are held entirely satisfactorily for long-time service. Yet, in the maple block-shear test (ASTM D 905-49) many of these adhesives have shear strengths of only 600 to 1000 psi.

Some adhesives for bonding flexible materials, such as fabrics or plastic films, are formulated to have plastic flow and to yield at rather low levels of strength in order to avoid excessive stress concentrations that might prematurely fail the adherends.

One important but largely unsatisfied need is reliable nondestructive test procedures for evaluating joint quality. Ideally, such a test should be sufficiently simple, rapid, and economical so that it could be used for screening every bonded product if the bond quality was considered critical. This type of test would then detect all bonds below a preselected standard of acceptance whether owing to a substandard adhesive sample or to improper bonding conditions. Because of the great importance of uniformly high quality bonds with structural metal-bonding adhesives for aircraft, considerable effort has been directed toward development and standardization of such nondestructive joint tests. Greatest emphasis has been on ultrasonic vibrations and instrumentation to detect responses from joints of different quality. Although considerable progress has been made and some industrial ultrasonic bond testers have been used in a few aircraft plants, it appears that at present such techniques are mainly suitable to detect a bond from a void. Their ability to distinguish a good bond from a mediocre or poor bond has not yet been adequately demonstrated.

Stability and working characteristics include storage life, viscosity or consistency, rate of spread, solids content, pH, flash point, and rate of strength development. The user may also require evaluation of such properties as degree of tackiness and tendency of one adhesive-coated surface to stick excessively to an adjacent one (blocking) (qv), or the detection by analytical methods of the presence of extenders, acids, or alkalies that are objectionable for certain uses. In some cases, it is necessary to make pilot runs on regular bonding equipment to determine the adequacy of the adhesive.

Because of the great strides in the development of synthetic-resin adhesives capable of long-time service under even severe service conditions, the need for adequate short-term tests of permanence is particularly great. A number of useful tests have been used to make crude comparisons of the relative permanence of different adhesives or of bonded joints. In order to give usable results within a few hours or days, the tests must involve rather severe treatments, such as boiling in water, steaming, heating, or exposure to heavy mold or other microbiological conditions. The adequacy of such tests can only be assessed by comparison with results of long-term controlled exposures or of actual service. Since service tests may require years before useful results are developed, the short-term tests must often be used before their validity can be fully established.   One present obstacle to development of more adequate permanence tests is a general lack of real understanding of the fundamental mechanisms by which adhesive bonds actually deteriorate in service.   When such mechanisms are more clearly understood, it will be relatively simple to accelerate these reactions for the short-term tests in which the user can place greater confidence.   In spite of these limitations, the potential user or purchaser has on hand a variety of test procedures for various adhesive applications and more are being actively investigated each year.

Bibliography

1. *ASTM Standard on Adhesives*, Philadelphia, Pa., 1957.
2. E. W. Kuenzi, *Determination of Mechanical Properties of Adhesives for Use in Design of Bonded Joints*, Report No. 1851, U.S. Forest Products Laboratory, 1956.
3. H. W. Eickner, *The Gluing Characteristics of 15 Species of Wood with Cold-Setting, Urea-Resin Glues*, Report No. 1342, U.S. Forest Products Laboratory, April 1955.
4. W. Z. Olson and R. F. Blomquist, "Polyvinyl-Resin Emulsion Woodworking Glues," *Forest Prod. J.* **5**, 219 (Aug. 1955).
5. H. W. Eickner, *General Survey of Data on the Reliability of Metal-Bonding Adhesive Processes*, Report No. 1862, U.S. Forest Products Laboratory, May 1957.
6. H. W. Eickner, *The Shear, Fatigue, Bend, Impact, and Long-Time Load Strength Properties of Structural Metal-to-Metal Adhesives in Bonds to 24S-T3 Aluminum Alloy*, Report No. 1836, U.S. Forest Products Laboratory, May 1959.
7. F. Werren and H. W. Eickner, "Climbing Peel Test for Strength of Adhesive Bonds," *Mod. Plastics* **34** (4), 187 (Dec. 1956).

R. F. Blomquist
Forest Products Laboratory
U.S. Department of Agriculture

ANALYSIS OF ADHESIVES

An introduction to the subject of adhesion and adhesives was given in the preceding article, ADHESION. The reader is referred to that article and the general references included therewith for background information.

Adhesive formulations are usually complex mixtures containing a polymer or resin, either natural or synthetic. Along with the resin may be found solvents, stabilizers, fillers, pigments, plasticizers, modifiers, lubricants, hardeners, and residual monomers. Partly because of the complex nature of adhesives, analyses at the consumer level are primarily concerned with quality and uniformity determinations. The actual composition of the adhesive formula is of much less value to the consumer and is seldom determined.

Evaluation of Adhesives

The first step in the evaluation of an adhesive is to determine by physical and mechanical tests that it is suitable for the intended application. Simple and reliable tests are then used to insure that subsequent shipments are of suitable quality and uniformity.

The evaluation methods described here are representative for various industries, but many modifications are encountered. Some lesser known specialty adhesives are not covered. Many of the polymers are used as adhesives without supplementary additives or are compounded by the consumer just prior to use.

Many of the tests given are of an empirical nature. For most meaningful results a comparison to a well-performing adhesive formulation is recommended.

Amino Resin Adhesives. The amino resins used as adhesives are usually of the urea–formaldehyde, melamine–formaldehyde, or urea–melamine–formaldehyde type. The urea–formaldehyde type is often supplied in a modified form. The most commonly encountered modifiers are furfuryl alcohols or polyvinyl acetates. The resins are supplied as solids or as liquid suspensions.

As with other solvent-borne adhesives, the *solids content* of a liquid amino resin adhesive is of prime importance. The solids determination for the amino resin adhesives most often encountered is the Plastics Materials Manufacturing Association (PMMA) *nonvolatile method* (1). Sufficient sample to contain 0.4–0.5 g of resin solids is weighed into a dried and weighed aluminum dish 57–58 mm in diameter; 5 ml of water is pipetted into the dish and mixed thoroughly by gentle rotation. The dish is placed in a mechanical convection oven at 105°C. After a 3-hr drying period, the dish is cooled in a desiccator containing calcium chloride or calcium sulfate desiccant and weighed rapidly. The % solids is calculated.

The *free-formaldehyde* content of amino-resin adhesives is most often determined by the acidimetric sulfite method (2). A 2–5-g sample is weighed into 100 ml of water in a 250-ml Erlenmeyer flask. If the sample does not dissolve completely, sufficient ethanol is added to effect solution. The mixture is cooled to 0–4°C in an ice bath. Twenty-five milliliters of M sodium sulfite is neutralized to the thymolphthalein end point and transferred to a 100-ml beaker. A 10-ml portion of N hydrochloric acid is added, and the mixture is cooled in an ice bath. The solution is titrated to the faint blue end point of thymophthalein with either 0.1 N sodium hydroxide or 0.1 N hydrochloric acid. The acid sulfite solution is added to the sample solution and the beaker is rinsed several times with cold water. The solution is then titrated rapidly with 0.1 N sodium hydroxide. The amount of free formaldehyde present is calculated from the amount of hydrochloric acid consumed in the reaction.

For the determination of *viscosity*, the most commonly used method is the Gardner-Holdt bubble tube method (3). A standard tube is filled with the resin to a prescribed mark. The time for the bubble to rise in the tube filled with the resin is compared with that for the standard liquids in the same size tube. The measurements are carried out at 25°C.

The determination of *pH* is also of the utmost importance. At times, the pH is determined in conjunction with a buffer curve (4).

Other tests used less frequently are the determinations of total formaldehyde, total nitrogen, total amine constituents (urea and/or melamine), total alcohol, total water, and ash. See also AMINO RESINS; MELAMINE RESINS; UREA–FORMALDEHYDE RESINS.

Animal Glues (Collagen). This natural high polymer is a protein constituent derived from bones, connective tissues, and skin. The glues are often evaluated in accordance with the methods adopted by the National Association of Glue Manufacturers (5,6).

To determine the *viscosity* a molten solution of the glue is prepared by dissolving 15 g of dry glue in 105 g of water. The viscosity of a 100-ml portion is measured at 60°C in a modified Bloom viscosity pipet. A modified TAPPI method (7) calls for a solution of any desired concentration at 60°C and for the use of a Brookfield viscometer at 20 rpm with the spindle that gives readings between 20 and 80% of the range of the scale.

Jelly strength is defined as the force required to depress the surface of a jelled solution of glue by 4 mm when measured by the Bloom gelometer (5). It is determined on the solution used for viscosity determination after it is jelled for 17 hr in a constant temperature bath maintained at 10 ± 0.1°C.

To determine the *moisture content*, the glue is ground to 8 mesh and dried for 17 hr at 105°C. The loss in weight is calculated as the percent moisture content. The *pH* is determined electrometrically on the melted glue jelly at 40°C. For the *foam* characteristics, a 12.5% solution at 140°F is agitated at high speeds with a milk shaker stirrer for 15 sec and the time for the foam to disappear is recorded. *Flow point* (7) tests are less often used. See also GLUES.

Blood and Soybean Glues. The primary adhesive component in these two glues is the protein. The commercially available product is usually the dry adhesive which contains only part of the ingredients required for the final adhesive. The quality control thus is similar to the analysis of the base materials. The *total protein* present is determined by the Kjeldahl nitrogen. Standard methods are used for the determination of *moisture* and *pH*. *Solubility* and *alkaline viscosity* behavior are often determined.

Because of on-the-spot preparation of adhesives mixes, the final criteria for this suitability and quality is the *mix viscosity* after the mix is prepared in accordance with the manufacturers' specifications. The common viscometers used are the Mac-Michael, Stormer, and Brookfield (8,9). The rate of viscosity change of a glue mix with time is often determined. See also Glues.

Caseins. Casein, the main protein in milk, was recognized early as an adhesive. It distinguishes itself from other glues of animal origin by forming a relatively water-resistant bond. Casein glues contain oils, fillers, viscosity-adjusting agents, preservatives, and plasticizers in addition to their main constituent, casein. Casein itself may be analyzed for particle size, moisture content, extraneous matter, alkali requirements, solution viscosity, nitrogen content, fat, acidity, and ash (10,11). See also Casein; Glues.

The final criterion for quality of the finished glue mix is the viscosity and the working life. For *viscosity* determinations, the Stormer viscometer with a 500-g weight and the Brookfield viscometer are the most generally used types. The approximate *working life* (12) of a glue mix is determined by observation of the change in spreading characteristics. The end of the working life has been reached when the adhesive can no longer be spread conveniently to produce a uniform film. For a more accurate test the viscosity of the glue mix is determined with a suitable viscometer at convenient time intervals of 30 min. An arbitrary end point of 800 P is chosen as the working life in some applications.

Celluosics. The cellulose derivatives form an important segment of the adhesive industry. They are used as adhesives in their own right in addition to being used as additives in many other glues. The most important aspect in their evaluation as adhesives is the viscosity characteristics of cellulose derivatives. Numerous other analyses are used which are similar to those for analyses of the base polymers and are specific for the particular derivative. See Cellulose; Cellulose derivatives. A good proportion of the testing is covered by the ASTM test methods for soluble cellulose nitrates (13,14), cellulose acetates (14,15), cellulose acetate butyrates (16), methylcelluloses (17), and ethylcelluloses (18).

Epoxy Resin Adhesives. Epoxy resins are relative newcomers to adhesive technology. Because they are usually supplied as two-component systems, the resin and the hardener or catalyst, the analysis of the polymer portion is similar to the analysis of epoxy resin.

One of the properties of epoxy resins which is of great importance is the *epoxide content*. The most common of the various procedures for its determination is the hydrogen bromide in acetic acid method (19). A 0.3–0.6-g sample of epoxy resin is weighed into a 50-ml Erlenmeyer flask and is dissolved in chlorobenzene or other suitable solvent; five drops of 0.1% crystal violet indicator are added. A moisture-protected buret is lowered into position and the sample is titrated to a blue-green end point with 0.1 N standardized hydrogen bromide solution in acetic acid. The epoxide equivalent per 100 g of resin is calculated.

In many applications the amount of *free chloride may not exceed certain levels*. The Volhard procedure is the preferred method for the determination of free chloride. The total chlorine is often determined by hydrolyzing the chlorohydrin groups in alcoholic potassium hydroxide for 1 hr at room temperature. The mixture is then acidified and the Volhard chloride determination is carried out. The difference between the total chlorine and the free chloride values gives the chlorohydrin content (20).

An often used quality control method for solid epoxy resin adhesives is *melting point* determination. The Durrans' mercury method is used in this determination (21). Other methods used include hydroxyl content, glycol content, and color determinations.

Fish Glues. Very little written technical information is available about glues prepared from fish skins. The glues are checked for their gel point, salt concentration, and viscosity by means of the Stormer viscometer. Fish glues can be found alone or mixed with animal glues, dextrin adhesives, polyvinyl acetate emulsions, and latex emulsions (22). See also GLUES.

Furan Resin Adhesives. The furan resins prepared from furfuryl alcohol or from furfuryl alcohol and formaldehyde are comparatively recent and do not find wide commercial application. Tests reported for resin-uniformity determination are moisture content, molecular weight, free furfuryl alcohol, saponification number, free formaldehyde, and buffer equivalent (23).

Hot-Melt Adhesives. The term *hot-melt adhesives* refers to a group of bonding agents which are applied in a hot liquid state and achieve a solid state and the resulting bonding strength on cooling. Their importance has increased in recent years because their rapid setting times make them particularly adaptable to use in high-speed, automatic equipment.

The hot melts include rosin and derivatives, mineral, vegetable, and petroleum waxes, coumarone–indene resins, polyamide resins, terpene resins, alkyds, thermoplastic phenol–formaldehyde resins often modified with ethylcellulose, polyvinyl acetates, butyl methacrylates, polyethylenes, polystyrenes, and other polymers. Hot melts also can include plasticizers, fluxing agents, fillers, pigments, dyes, and stabilizers. The generally applicable tests for the hot-melt adhesives include temperature and viscosity characteristics, heat stability, and softening-point determinations.

For *temperature* and *viscosity* characteristics (24) determination, a pint can is filled threé quarter's full with the adhesive to be tested. The can is fitted with a cover which has holes for a Brookfield viscosity spindle and for a thermometer. The adhesive is heated with a suitable outside heater to slightly above the desired maximum temperature. The desired temperature depends on the intended use and melting point of the adhesive. The viscometer spindle is placed in position and readings are taken at 10°F intervals as the system cools.

The *heat stability* (24) test consists of filling a pint can to four fifths of its capacity with the hot-melt adhesive. The can is placed into an electrically heated wax bath which has been adjusted to the temperature at which the hot melt will be applied in practice. The mix is rapidly stirred. Brookfield viscosity measurements are made every 4 hr or as experience requires. The test is considered complete when the hot melt gels, or at the end of 72 hr if the product remains stable for that length of time. Concurrently, a static test is run in two glass jars, one for separation, skinning, discoloration, and gelation, the other for viscosity comparison.

To determine *softening point* ring and ball and TAPPI methods are used, (25, 26).

Phenolic and Resorcinol Resin Adhesives. The most common adhesive of this class is the straight phenol–formaldehyde resin. Testing methods for quality of the different types in this class are quite similar.

Various methods for determining *nonvolatiles* are in use. They all give only empirical values, and for reproducible results the methods should be followed closely. One of the common methods is the oven solids method of the West Coast Adhesives Manufacturers (27). One gram of resin is weighed on an aluminum dish so it forms a small button. The sample is dried in a mechanical convection oven at 125°C for 105 min. The residual weight of the sample is determined after cooling in a desiccator and the percent nonvolatile matter is calculated.

In alkali-catalyzed phenolic resin adhesives, the amount of *total alkali* is determined. A potentiometric method has proven to be very convenient (28). A 2-g resin sample is weighed into a 250-ml beaker and is diluted with 10 ml of water. After the glass/calomel electrodes have been introduced, the mix is slowly titrated with standardized 0.1 *N* hydrochloric acid solution. The mix is kept well agitated with a magnetic stirrer. If cloudiness develops during the titration, isopropyl alcohol is added, up to 50 ml, to keep the resin in solution. The titration is continued to pH 4. The milliequivalent of acid used is the total basicity of the resin. The solution may be saved for free formaldehyde determination.

Free formaldehyde can be determined on the above solution after alkali determination or on a fresh solution adjusted to pH 4 (28). Ten ml of 10% hydroxylamine hydrochloride solution previously adjusted to pH 4 is added to the selected sample. The mixture is titrated with 0.1 *N* standardized sodium hydroxide solution 5 min after the addition. The milliequivalent of sodium hydroxide used gives the milliequivalent of free formaldehyde present.

Any free *phenol* must be separated prior to its determination. A suitable method for separation is steam distillation.

Viscosity is usually determined by the Gardner-Holdt bubble tube method (3). As a reference method, use of the Stormer viscometer is recommended (29).

Other tests commonly required are refractive index, specific gravity, hot plate cure time, gel time, and ash determination.

Polyamide Resin Adhesives. This group consists of two types of polyamide resins. One is the linear polymer made from acid and diamine. It is an unreactive thermoplastic material and is used as a hot-melt adhesive. The second is a thermosetting adhesive made from acid and polyamine. This polymer is usually highly branched and often finds use as a curing agent in epoxy adhesives.

The thermosetting polyamide adhesives are often marketed as solutions. The tests used to determine solution viscosity and solids content are the same as those used for amino resins adhesives. For better characterization, the carboxyl and amine end groups are sometimes determined (30).

Polyester Adhesives. This class of resins is not widely used as an adhesive. It is used at some specialty applications or as an adhesive modifier. It can be often found used in conjunction with rubber and phenolic resin adhesives. For analysis, see POLYESTER RESINS.

Rubber-Based Adhesives. Rubber-based adhesives include a wide variety of materials ranging from natural rubber to modified synthetic polymers. The composition of rubber-based adhesives could include some or all of the following materials: elastomers, tackifiers, fillers, plasticizers, antioxidants, curing agents, and sequestering agents. With the wide variety of available materials, no specific analysis scheme is available. The more generally applicable tests are discussed below and others have been described (31,32).

For the determination of *total solids*, samples up to 10 g are oven dried at 70°C to constant weight or for 16 hr. Higher temperatures are not recommended as skinning might take place or volatile adhesive components might be removed.

For the *viscosity* determination, various types of viscometers are in use, such as Saybolt, Zahn, Stormer, Brookfield, and flow-through capillaries.

The latex-type rubber-based adhesive is tested for *mechanical stability*. The latex is diluted to 55% total solids content and stirred with a high-speed stirrer (14,000 rpm) until visible clots are formed.

Soluble Silicates. This class of adhesives is important due to low cost and formation of totally combustion-resistant glues. The commercial liquid sodium silicates consist of silicon dioxide and sodium oxide. Modifications include starches, proteins, aqueous latexes, polyvinyl acetates, methylol ureas, and clays.

The most important determination for this type of adhesive is the *alkali-to-silicate ratio*, since this determines the adhesive property of the sample (33,34). The alkali is determined as *sodium oxide* by titration with standardized hydrochloric acid solution. Concentrated hydrochloric acid is added to the solution from the sodium oxide determination and the solution is evaporated to dryness. The residue is dissolved in 1:1 hydrochloric acid and the mixture is digested on a steam bath for 10 min. After digestion, the mixture is filtered through ashless filter paper. Concentrated hydrochloric acid is added to the filtrate and the mixture is evaporated to dryness. The residue is transferred to the same filter paper and washed free from acids. The paper and the residue are ignited in a platinum crucible in a muffle furnace until free from carbon by slowly increasing the temperature to 1000°C. After cooling, the residue is moistened with 1:1 sulfuric acid and the crucible is heated on a hot plate until fumes of sulfur trioxide appear and then over a burner until the fumes disappear. Finally, the crucible is ignited in a muffle furnace at 1000°C for 30 min. After cooling in a desiccator, the crucible is weighed to the nearest milligram. The *silicon dioxide* content is then determined by loss in weight on treatment with hydrofluoric acid; see SILICON.

Viscosity is measured by the Stormer viscometer with a special silicate test cup without the thermometer well and central baffle plate. For *specific gravity* measurements, a Baumé hydrometer or gravimetric method is used.

Silicone Resin Adhesives. These resins have found commercial application only recently. They are noted for their excellent water resistance and applicability to relatively high-temperature use conditions.

The principal property of interest to the adhesive chemist is the amount of resin present in solution. The common nonvolatile determinations are not applicable due to the high volatility of lower molecular weight silicones. A good method which has found general application in adhesion is the colorimetric determination of silicon (35). The sample is fused with sodium peroxide in a Parr-type bomb. A silicon molybdate complex is formed by reaction with ammonium molybdate. The complex is reduced with sodium sulfite to the heteropoly blue. The blue color is measured with a suitable spectrophotometer and the amount of silicone is calculated by the use of a standard curve; see SILICONE RESINS. Some of the ASTM tests could be used (36).

Starch and Dextrin Adhesives. The principal component of starch adhesives is the polymer of dextrose synthesized by the plants. Tests used for the raw material are also often applicable to the adhesive and routinely include determination of acidity and alkalinity, cold water solubles, moisture, ash, pH, starch origin by microscopic examination, and protein content by nitrogen determination (37,38). Specific tests to determine the suitability of starch adhesives have been described (38–40).

The amount of *reducing sugars* is determined by Schoorl's method involving oxidation with Fehling's solution and titration with standardized 0.1 N sodium thiosulfate.

To determine *alkali fluidity*, a 5-g sample is weighed into a 300-ml beaker and 10 ml of water, at 25°C, is added. The mixture is slurried and placed into a 25°C constant temperature bath; 90 ml of 1% sodium hydroxide solution at 25°C is added and the mixture is stirred at 200 rpm for 3 min and allowed to stand in the bath for a total of 32.5 min. Then the mixture is transferred to a standard fluidity funnel. At 33 min total elapsed time, the paste is allowed to flow into a graduated cylinder. The volume that will flow during a 70-sec interval is the measure of fluidity.

Cold water solubles are determined by adding water to a 50-g sample to give a total volume of 250 ml. After 30 min of shaking, the suspension is filtered through a Whatman #12 filter paper. Then 100 ml of filtrate is transferred to a weighed evaporating dish, evaporated to dryness, and dried in a vacuum oven for 1 hr at 100°C. After cooling, the residue is weighed and the percent solubles is calculated.

The *solids* are determined by the usual methods. For the *viscosity* determination, either the Brookfield or Stormer viscometer can be used. *Acidity* or *alkalinity* is determined either by titrimetry or by measurement of pH. *Color* is determined either on the liquid adhesive or on a dried film by comparing it against a series of standards.

Vinyl Resin Adhesives. This classification includes a large number of homo- and copolymers. In a sense, any polymer formed from a monomer containing the unsaturated double bond is included.

Viscosity is determined on dilute solutions of the polymers by a capillary viscometer. The ASTM method (41) is applicable with the proper choice of solvents.

Various methods are used to determine the amount of *nonvolatile matter*. A general method (42) for vinyl adhesives calls for weighing about 3–5 g of sample into a tared aluminum dish. The dish is placed in an oven at 105°C for 2 hr. After the dish is removed and cooled in a desiccator, the contents are weighed. The percent residue is calculated as nonvolatile matter.

The *specific gravity* and the *particle size* are also often determined (43,44).

Separation and Identification

An analyst will encounter many difficulties in attempting to determine the composition of various adhesives. The problem of distinguishing among large numbers of available adhesives is compounded by the great variation in their compositions.

To analyze the adhesive and/or additives successfully, the analyst is usually confronted with the necessity for some degree of separation of the various components. In the majority of the methods available, the separation of the base adhesive from the components is dependent on differential solubilities. A suitable solvent for extraction must be found largely by trial and error.

Many heat-insensitive adhesives can be separated by the following method (45). A sufficient amount of adhesive for analysis is mixed with a solvent, preferably low-boiling, in which the polymer portion is soluble so as to obtain solution concentrations of 1–2% of polymer. The solution is passed several times through a glass filter to remove insoluble components, such as fillers and pigments. The polymer is precipitated at an elevated temperature with an appropriate precipitant (acid, base, or nonsolvent). The precipitate is separated from the supernatant liquid by decantation, filtration, or centrifugation and washed with fresh, hot precipitant. The precipitate is then dissolved in another solvent and precipitated with a low-boiling precipitant. The precipitate is washed and the procedure is repeated. Finally, the precipitate is dried at room temperature under vacuum. Repeated precipitation is usually, but not always, necessary.

For the thermosetting resins the elevated temperatures must be avoided. Limited success is possible by modifying the method above and making the separations at room temperature or lower (46). Further modification of the base adhesive to form heat-insensitive compounds and the use of higher temperatures may be helpful. Hydrolysis and esterification are examples of reactions which may be used to modify the basic adhesive.

A method for separation of polymers from solvents and monomers by a freezing technique is often quite applicable to adhesive components (47).

Gravity separation methods are at times used for the protein-type glue mixes. The following procedure has been found quite successful (28). To 100 g of glue in a 600-ml beaker, 400 ml of carbon tetrachloride is added. The slurry is stirred gently for 5 min. The mix is allowed to stand for about 30 min or until the inorganic chemicals settle to the bottom and the proteins and fillers rise to the top. The layers are separated. The solid phases are washed with carbon tetrachloride, all of which is collected and evaporated on a steam bath. The residue remaining after evaporation is usually the antifoam. The bottom layer, composed of inorganic chemicals, is dried and used for qualitative and quantitative tests for components. The top layer is transferred to a 600-ml beaker and 150 ml of carbon tetrachloride is added; 50-ml increments of benzene are added to this slurry. The slurry is stirred and allowed to separate after every addition. This addition is continued until a clean separation between protein (top) and other components (bottom) is obtained. This is usually obtained at 1:1 benzene/carbon tetrachloride ratio. The layers are separated and the solvent is removed by filtration. For better selectivity, the procedure is repeated. At times, further separation of the bottom layer can be achieved by the use of screens because of the differences in particle size.

Once the separation has been obtained, the identification of the separated compounds can be made in accordance with any conventional scheme. The most comprehensive scheme published for polymer identification can be found in Kline (48,49), which is applicable for any polymer encountered in the adhesive chemistry. Additional helpful hints are given by Lucchesi (50). Many of the described spot and color tests can be used on the sample directly without prior separation; however, extreme caution must be exercised to avoid erroneous conclusions.

Bibliography

1. *Procedure for the Determination of Total Solids in Liquid Urea Adhesives*, Plastics Materials Manufacturing Association, revised by Manufacturing Chemists' Association, Washington, D. C., 1952.
2. P. R. Averell, "Amino Resins," in G. M. Kline, ed., *Analytical Chemistry of Polymers*, Part I, Interscience Publishers, Inc., New York, 1959, p. 63.
3. *ASTM D 154–65, Methods of Testing Varnishes*, American Society for Testing and Materials, Philadelphia, 1965.
4. Ref. 2, pp. 88–89.
5. National Association of Glue Manufacturers, "Standard Methods (revised) for Determining Viscosity and Jelly Strength of Glue," *Anal. Chem.* **2,** 348 (1930).

6. J. R. Hubbard, "Animal Glues," in I. Skeist, ed., *Handbook of Adhesives*, Reinhold Publishing Corp., New York, 1962, p. 118.

7. J. P. Richards, "Animal Glue Base Adhesives," in W. H. Neuss, ed., *Testing of Adhesives*, Technical Association of the Pulp and Paper Industries, New York, 1963, pp. 42–44.

8. A. L. Lambuth, "Soybean Glues," in I. Skeist, ed., *Handbook of Adhesives*, Reinhold Publishing Corp., New York, 1962, p. 155.

9. A. L. Lambuth, "Blood Glues," in I. Skeist, ed., *Handbook of Adhesives*, Reinhold Publishing Corp., New York, 1962, pp. 167–168.

10. E. Sutermeister and F. L. Browne, *Casein and Its Industrial Applications*, 2nd ed., Reinhold Publishing Corp., New York, 1939, pp. 149–168.

11. *TAPPI Standards and Suggested Methods T 607 ISU-66*, Technical Association of the Pulp and Paper Industries, New York, 1966.

12. U. S. General Service Administration, "Adhesives, Casein-Type, Water and Mold Resistant," *Federal Spec. MMM-A-125a* (1955).

13. *ASTM D 301-56, Specification and Methods of Test for Soluble Cellulose Nitrate*, American Society for Testing and Materials, Philadelphia, 1965.

14. *ASTM D 1343-56, Method of Test for Viscosity of Cellulose Derivatives by Ball Drop Method*, American Society for Testing and Materials, Philadelphia, 1965.

15. *ASTM D 871-63, Methods of Testing Cellulose Acetate*, American Society for Testing and Materials, Philadelphia, 1965.

16. *ASTM D 817-65, Methods of Testing Cellulose Acetate Propionate and Cellulose Acetate Butyrate*, American Society for Testing and Materials, Philadelphia, 1965.

17. *ASTM D 1347-64, Methods of Testing Methylcellulose*, American Society for Testing and Materials, Philadelphia, 1965.

18. *ASTM D 914-50, Methods of Testing Ethylcellulose*, American Society for Testing and Materials, Philadelphia, 1965.

19. A. J. Durbetaki, *Anal. Chem.* **28**, 2000 (1956).

20. R. E. Burge, Jr., and B. P. Geyer, "Epoxy Resins," in G. M. Kline, ed., *Analytical Chemistry of Polymers*, Part I, Interscience Publishers, Inc., New York, 1959, pp. 140–141.

21. H. A. Gardner and G. G. Sward, *Physical and Chemical Examination of Paints, Varnishes, Lacquers and Colors*, 11th ed., Henry A. Gardner Laboratory, Bethesda, Md., 1950, p. 468.

22. H. C. Walsh, "Fish Glue," in I. Skeist, ed., *Handbook of Adhesives*, Reinhold Publishing Corp., New York, 1962, pp. 126–127.

23. L. H. Brown, "Furan Resins," in G. M. Kline, ed., *Analytical Chemistry of Polymers*, Part I, Interscience Publishers, Inc., New York, 1959, pp. 206–218

24. P. H. Yoder and D. D. Dooley, "Hot Melt Adhesives," in W. H. Neuss, ed., *Testing of Adhesives*, Technical Association of the Pulp and Paper Industries, New York, 1963, pp. 92–94.

25. *ASTM E 28-58T, Method of Test for Softening Point by Ring and Ball Apparatus*, American Society for Testing and Materials, Philadelphia, 1965.

26. *TAPPI Standards and Suggested Methods T 630 m-61*, Technical Association of the Pulp and Paper Industries, New York, 1961.

27. *Test Method #2.2*, Technical Committee West Coast Adhesives Manufacturers Association, Seattle, Wash., Aug. 23, 1957.

28. Weyerhaeuser Company laboratories, unpublished data.

29. *Test Method #1.1*, Technical Committee West Coast Adhesive Manufacturers' Association, Seattle, Wash.

30. J. E. Waltz, *Anal. Chem.* **19**, 448 (1947).

31. F. Fisher, "Latex-Based Adhesives," in W. H. Neuss, ed., *Testing of Adhesives*, Technical Association of the Pulp and Paper Industries, New York, 1963, pp. 95–106.

32. *1964 Book of ASTM Standards*, Part 28, American Society for Testing and Materials, Philadelphia, 1964.

33. J. G. Vail, *Soluble Silicates, Their Properties and Uses*, Vol. 1, Reinhold Publishing Corp., New York, 1952, pp. 33–43.

34. *TAPPI Standards and Suggested Methods T 632 ts-63*, Technical Association of the Pulp and Paper Industries, New York, 1963.

35. H. L. Kahler, *Anal. Chem.* **13**, 536 (1941).

36. *ASTM D 1346-66, Methods of Testing Silicone Insulating Varnishes*, American Society for Testing and Materials, Philadelphia, 1966.

37. R. W. Kerr, *Chemistry and Industry of Starch*, 2nd ed., Academic Press, New York, 1950.
38. *TAPPI Standards and Suggested Methods T 638 m-53*, Technical Association of the Pulp and Paper Industries, New York.
39. E. Bearman et al., "Starch and Dextrine Adhesives," in W. H. Neuss, ed., *Testing of Adhesives*, Technical Association of the Pulp and Paper Industries, New York, 1963, pp. 7–10.
40. *Manual of Testing Methods for Labeling Adhesives*, The Packaging Institute, New York, 1958.
41. *ASTM D 1243-60, Methods of Test for Dilute Solution Viscosity of Vinyl Chloride Polymers*, American Society for Testing and Materials, Philadelphia, 1965.
42. C. J. Kennett, "Vinyl Polymers and Copolymers," in G. M. Kline, ed., *Analytical Chemistry of Polymers*, Part I, Interscience Publishers, Inc., New York, 1959 pp. 449–488.
43. *Ibid.*, pp. 466–470.
44. *ASTM D 792-60 T, Methods of Test for Specific Gravity and Density of Plastics by Displacement*, American Society for Testing and Materials, Philadelphia, 1965.
45. H. Mark, *Anal. Chem.* **20,** 104 (1948).
46. T. P. G. Shaw, *Anal. Chem.* **16,** 541 (1944).
47. F. M. Lewis and F. R. Mayo, *Anal. Chem.* **17,** 134 (1945).
48. G. M. Brauer and E. Horowitz, "Systematic Procedures," in G. M. Kline, ed., *Analytical Chemistry of Polymers*, Part III, Interscience Publishers, Inc., New York, 1959, pp. 1–140.
49. G. M. Brauer and S. B. Newman, "Color Tests," in G. M. Kline, ed., *Analytical Chemistry of Polymers*, Part III, Interscience Publishers, Inc., New York, 1959, pp. 141–259.
50. C. A. Lucchesi and D. J. Tessari, *Offic. Dig. Fed. Soc. Paint Technol.* **34,** 387 (1962).

EINO MOKS
Weyerhaeuser Company

BONDING

The art of bonding has been practiced by man for many thousands of years, some of the earliest recorded applications being the Egyptian murals and veneered caskets in our museums. The ancient art of gluing, as reviewed by Darrow (1), is essentially the technique of application of naturally occurring high polymers as adhesives. These range from blood- and bone-derived animal glues, fish glues, natural asphaltic materials, natural tree gums, etc, to the more recent and relatively more sophisticated compositions based on vegetable starches and natural rubber latex.

Certain living animals are expert in performing polymeric processes, such as the production of polyamide fiber by the silkworm and the molding of a chitinous shell by many insects. We also note interesting polymeric adhesive developments in nature. Among these are the adhesives employed by molluscs for attachments or for bonding their living muscles to their calcareous shells, by many spiders for coating their webs, on the footpads of the common house fly, and in the binder compound used in the reinforcing of birds' nests.

As practiced by modern man, bonding with adhesives has become a widely accepted, important production tool, and the manufacture of adhesive materials is an important and growing segment of the chemical industry.

Joining Methods for Plastics

The ability of plastic materials to be formed, cast, or molded homogeneously, and frequently anisotropically, into a great variety of shapes is one of their greatest assets. Although many articles of complex contour may be molded in one piece, nevertheless the commercial utilization of plastics has been greatly enhanced by the successful application of several bonding processes, resulting in composite articles including plastic parts. These articles may consist of assemblies of various stock or special shapes of the same plastic; they may be combinations of different plastics; or they may consist of plastics bonded to other materials. In these latter cases, the assembly may be designed to benefit from the best chemical, structural, or surface characteristics of each of the components. There are four basic methods of joining plastics.

Solvent Cementing (also known as solvent welding). The surfaces of the plastic materials to be bonded are brought to a fluid or tacky condition by the application of a solvent for the plastic. In special cases the solvent may be a catalyzed monomer, or it may contain dissolved polymer, the same as that of the plastic part. The resultant joint is a cohesive joint consisting essentially of the parent material and having joint properties approaching the properties of a homogeneous piece. This process is most suitable for the more amorphous and soluble thermoplastics, such as acrylic resins, cellulosic resins, and polystyrenes. It is not generally useful for the more crystalline, insoluble thermoplastics such as polyethylene, nylon, and the fluorocarbon plastics, and cannot be used at all for thermoset plastics. Dissimilar thermoplastics may be solvent cemented, providing that they are compatible with each other in solution and in a molten condition.

Adhesive Bonding. In this process a layer of an adhesive composition chemically different from the plastics to be bonded is interposed between the adherends. The properties of the resultant joint depend upon the specific adhesion characteristics of the materials, the cohesive strength of the adhesive, and the design of the joint. Adhesive bonding is a method that is suitable for all plastics, thermosetting as well as thermoplastic. In many instances, the bonded joint may be stronger than the plastic adherends themselves.

Welding. The surfaces of the adherends are brought to a molten or tacky condition by the application of heat and are subsequently brought together, held, and cooled. This method is suitable for virtually all thermoplastics, although it is particularly useful for joining the more crystalline types such as poly(vinyl chloride), nylon, and polyethylene. Of course, it cannot be used for thermoset plastics. Since fusion may not be complete in all cases, the strength of the resultant joint may be less than that of the parent material, and oxidative or thermal degradation may further contribute toward weakness. Heat may be applied by many methods, such as gas or electric gun, heated tool, induction heating, friction (spinning), etc. (The use of an interposed layer of a molten material, if chemically different from the parent plastic adherends, is classified as an adhesive-bonding technique.)

Mechanical Fastening. Nails, rivets, grips, and similar mechanical fasteners provide a purely mechanical lock or frictional fit, at the specific locations of the fasteners. This technique is, of course, suitable for all plastic materials, depending upon the allowable stress configuration in the joint.

As compared with mechanical-fastening techniques, there are many advantages in the bonding and welding methods and many applications where mechanical fasteners may not be used at all.

1. Plastics are frequently used in all-plastic assemblies, where exposed metals must be avoided, such as in applications where acid resistance is required. This obviously precludes the use of metal screws or rivets, etc.

2. Plastics may be used in composite articles to cover or protect metals. Here an unbroken plastic surface is essential.

3. Mechanical fasteners introduce stress concentration, which must be carefully evaluated by the designer in regard to the strength of the components.

4. Adhesive bonding also is capable of joining dissimilar materials (as contrasted to the solvent-cementing and welding techniques above) and as compared with mechanical fasteners, provides uniform stress distribution over the entire bonded area, a smooth surface contour resulting from the elimination of projecting fasteners, reduced weight, and sealant, insulation, and vibration-damping properties.

The classification of plastics in accordance with their fusibility, crystallinity, and polarity (measured by solubility parameter) can be very useful in predicting their behavior with various adhesives and solvents and the preferable joining method (2). A study of these factors indicates the following: (*1*) The crystalline thermoplastics are best bonded by heat rather than adhesives (ie, the best adhesive is the polymer itself in molten form); the crystalline materials melt more sharply than the amorphous plastics, and this property can be quite useful in welding techniques. (*2*) Amorphous thermoplastics are readily bonded to themselves by solvent cementing. (*3*) Thermosetting plastics, which are inherently insoluble and infusible, must be bonded with adhesives, and do not lend themselves to solvent-cementing or heat-welding techniques. (*4*) The specific adhesion characteristics of polymeric adhesives required for composite assemblies can generally be predicted. (*5*) Solvents, polymers used as adhesives, and plasticizers should be selected to correspond to the adherends in solubility parameter; this may require the use of a two-layer adhesive if there is a substantial difference between the solubility parameters of the two adherends.

When heated above their melting points, the crystalline plastics become soluble, and can be made into cements, gel lacquers, or coating doughs. The melts may be used to bond the plastics to themselves or to nonplastics, through heat alone. The fundamental steps involved in the first two joining methods, solvent cementing and adhesive bonding, are: design and selection of the solvent or adhesive; design of the joint; surface preparation of the adherends; preparation of the adhesive; wetting the adherend surfaces with the solvent or adhesive; positioning the adherends and supporting the assembly during some finite period of time, during which the junction gains cohesive strength; and removing any artificial supports or devices and restoring the assembly to its designed ambient service conditions.

Selection of the Adhesive or Solvent

The first step in the bonding operation is the design, if necessary, and selection of the most suitable adhesive or solvent cement for use in the operation under consideration. See also Adhesive Compositions, p. 482.

Design and Selection of the Adhesive. Adhesive bonding implies the introduction between the adherends of a layer of polymeric adhesive, usually different in composition from the adherends. One widely accepted definition of an adhesive is "a substance capable of holding materials together by surface attachment" (3). From the practical point of view, this definition suffers by its broadness, since it does not require any significant degree of strength or permanence in the bond attained. A useful adhesive is one that has sufficiently high permanence, cohesive strength, specific adhesion, and resistance to the physical and chemical atmosphere in which the assembly will exist, to contribute significantly to the structural strength of the assembly throughout its useful life.

During the past twenty-five years, a revolution has occurred in the usefulness and in the adaptation of adhesives, due primarily to the introduction into adhesive compositions of synthetic resinous and elastomeric materials, as well as to better refinement and quality control of natural polymers. Adhesives are now available with a high degree of permanence, high cohesive strengths, confident reproducibility, versatility in specific adhesion, and many different physical as well as chemical forms permitting a variety of assembly processes.

The great advantages of the recently available synthetic polymeric compositions are that they exhibit structural integrity in adhesive-bonded assemblies over a wide range of atmospheric conditions and even with such strong, rigid adherends as metals and reinforced plastics. With plastics particularly, the general chemical relationships between synthetic adhesives and plastics not only contribute to good specific adhesion, but also assure that a close match exists with regard to modulus, flexibility, toughness, cohesive strength, and chemical resistance. These qualities have led, for example, to the development of reinforced and thermosetting plastics, whose great strength and usefulness under a wide range of environments and conditions have led to their use in many new applications, frequently in combination with metals and other rigid materials.

If the adhesive has been properly formulated and properly selected as to chemical composition, then it will exhibit not only adequate cohesive strength, but also adequate specific adhesion.

The key to the adhesive-bonding process is the understanding that the adhesive must in some way be made fluid so as to wet the surfaces, and subsequently must be rigidified by chemical or physical reactions to develop adequate internal strength and load-bearing ability. Also, virtually every adhesive, regardless of bonding mechanism or chemical type, passes through a period of low cohesive strength during which a bonding tool or fixture takes over the job of temporarily maintaining the structural integrity of the system.

With this basic principle in mind, bonding processes can logically and readily be designed to suit virtually all types of adhesive, whether classified chemically, such as "neoprene–phenolic contact-bond types," "epoxy–amine pastes," "polyamide hot melts," and thermoplastic or thermosetting; or by physical form, such as water dispersions or emulsions, monomeric resinous pastes, lump hot melts, powders, films, solvent cements, etc. All of the bonding steps fall logically into place. Reduction in viscosity is effected by solution, emulsification, melting in a pot, melting under pressure in a hydraulic press, or employing a supercooled liquid (such as pressure-sensitive adhesive). Typical viscosity-increasing steps include drying of water or solvents, cooling of a melt, polymerization of a monomer, crosslinking, even chemical reactions with the adherends themselves. In many cases, combinations of these procedures are required for a single adhesive.

Assuming that the cohesive strength of the adherend is high, and at least is not limiting, the highest overall bond strength for any particular joint design will result if the following three conditions are met: (*1*) adequate wetting of the adherends by the adhesive, provided by proper surface preparation and cleanliness of the adherends and by maintaining minimum viscosity of the adhesive at the time of application; (*2*) proper formulation for adequate specific adhesion and adequate inherent strength in the direction or directions in which the joint will require resistance to stress; and (*3*) maximum attainment of the inherent cohesive strength of the adhesive material, characterized by maximum viscosity in the finished joint whether due to freedom from retained solvent or other volatiles, or to increased molecular weight as the result of a chemical reaction.

The selection of a particular adhesive as the unifying component of a specific assembly is a decision that requires a thorough understanding of the end product, its components, and its performance characteristics. The adhesive is merely one component of the process of assembly of the product, and it must be designed to suit the product and to suit the plant facilities in which the product is being manufactured (see also under Adhesive Compositions).

There are no universal adhesives and there are no simple criteria to guide the layman. For example, the multiplicity of chemical families of raw materials available to the formulator and the ease with which a competent chemist can manipulate these materials so that adhesives of equivalent performance characteristics may be available from several different types of raw materials make selection of an adhesive by chemical type (eg, an epoxy, a neoprene) utterly fallacious. Virtually identical chemical compositions of active ingredients may be made available in a variety of physical forms, such as solvent-free pastes or films, solvent cements, water dispersions, etc. "Minor" variations in ingredients used to modify a film-forming polymeric base may substantially change specific adhesion, water resistance, or peel strength, as well as drying and curing times.

Fig. 1. Fast cold-setting wood adhesive based on thermosetting emulsion passes Type I "exterior" boil test. Courtesy *Building Products*.

The *design* of an adhesive composition for a particular end product requires a thorough understanding and consideration of at least the following factors:

Specific Adhesion. Although the formulator may gain some guidance by application of the theory of adhesion, knowledge regarding which classes of adhesives and which specific modifying ingredients provide adhesion to a particular adherend is essentially an empirical science, and is part of the experience and memory of the formulator. In this regard, adhesive compounding is an art.

Properties of the Adherends. The formulator must be advised as to a great many physical and chemical properties of the adherends; preferably, he should have physical samples of the adherends available for direct observation and for empirical confirmation of his recommendations. He must consider their heat sensitivity to establish an upper temperature limit on such factors as adhesive melting points, heat-seal temperatures, curing temperatures, etc; whether they will be dissolved, swelled, distorted, or crazed by water or by particular solvents; whether the adherends will be damaged by acid or alkaline ingredients in the adhesive; whether reactive groups on the adherends may catalyze or disturb the curing mechanism of the adhesive; whether they are fragile enough to distort or break if bonded at any particular pressure; if the surface is porous so that the viscosity of the adhesive may be designed to penetrate—or alternatively, to bridge—the discontinuities; the nature of the surfaces; and which surface preparation methods may be required.

Performance Characteristics. The design of the adhesive will also depend on what will be required of the assembly in initial strength and in percentage retention of strength, after a specified time exposure to such conditions as elevated or lowered ambient temperatures (Fig. 1), high-humidity exposure, salt atmosphere or salt spray, high or low pH, immersion in solvents or chemicals, repeated washing, etc. As important as the amount of physical stress that may be encountered is the type of stress, eg, tension, shear, peel, or cleavage. This is intimately related to the joint design.

Dimensions and Contour of Adherends. These will influence, and in many cases fix, the design of the joint and the type of equipment that must be used for applying

THANK YOU for submitting your adhesive product inquiry.   Your cooperation in completing this questionnaire fully will assist our technical personnel in serving you effectively. PLEASE BE EXPLICIT and DETAILED. Use the back of this form if necessary. SUBMIT SAMPLES (under separate cover) of materials to be processed, other adhesives for comparison, sketches, purchase specifications, etc.

1. Your finished product_____
   Component materials_____
   _____
   Joint configuration_____
   _____
   Description of process_____
   _____

2. Your application equipment_____
   _____
   Your drying equipment_____
   _____Time/temperature limits_____
   Your curing equipment_____
   _____Time/temperature limits_____
   Pressure_____Other_____

3. Service requirements (specify strength, weathering, service temperature, washability, etc.)
   _____
   _____

4. Other characteristics: Color_____Viscosity_____
   MUST be one-part ☐   MUST be non-inflammable ☐   MUST be 100% solids ☐ film ☐
   paste ☐   powder ☐   hot melt ☐   other_____

5. Economic justification: Estimated volume (lb, gal, ft² of bonded area)
   _____per_____Seasonal ☐ if so, when_____
   Price or cost limitation_____Is this a new use of adhesives
   ☐ yes   ☐ no; If no, what is the competitive product used now?_____
   _____Where does it need improvement?_____
   Other comments_____
   _____
   _____

Fig. 2.   Adhesives selection guide.   Courtesy Talon Adhesives Corporation.

the adhesive and as bonding fixtures during the process steps.   Obviously, different procedures and probably different adhesives must be used for continuous lamination of films, in contrast to the bonding of small rigid parts, and again for the bonding of rigid parts with a large surface area.   For coarsely fitting parts, a void-filling adhesive will be required; sometimes it may be required to foam.   If at all feasible, the joint should be designed in accordance with the known good practice rules for adhesive assemblies; whatever the eventual shape of the joint, it will influence the particular strength characteristics that must be made available by the adhesive.

Plant Facilities.   The working conditions and the existing bonding tools (or the budget for making available new tools) of the plant in which the adhesive assembly is to be made may be limiting in regard to the bonding process which can be employed. For example, if ventilation is inadequate or if open flames or sparking electrical equipment surround the assembly area and cannot be removed, then the formulator must avoid solvent-dispersed adhesives of the inflammable or toxic type. The bonding process must frequently be designed to fit existing equipment; and accordingly available temperatures, pressures, oven-drying capacities, time–temperature relationships, etc, must be thoroughly understood. These characteristics will affect the form of the adhesive (whether it should be a fluid cement or water dispersion, a hot melt, etc) and also the drying rate of the solvent or the curing rate of the resin system. Within the maximum flexibility of design available in the plant, the formulator should provide his product in the form that results in the simplest bonding process (least number of steps and least amount of machinery requiring trained personnel or careful control), the safest process (avoiding inflammable or toxic materials if at all possible), and the most economical process.

It is thus apparent that the design of an adhesive is inseparable from the design of the bonding process for the product, and is accordingly a joint venture between the adhesive formulator and the product engineer in the user's plant; cooperation between these two individuals is required. In order for an adhesive salesman to represent the formulator effectively, he must have adequate technical training; similarly, in order for a purchasing agent to represent the design engineer, he must have adequate understanding of his company's product plans and production facilities.

A checklist is employed by many adhesive companies to assure that all of the information necessary to understanding the adhesive design is provided to the formulator (Fig. 2). After study of this form, the formulator should be able to produce for his customer's evaluation an adhesive sample that may require only minor subsequent modification, if any.

Only after the adhesive companies have had an opportunity to design their adhesive to the indicated product and process specifications does *selection* become meaningful. Selection of an adhesive by the end user should be from among competitive adhesives presumably designed to equivalent specifications, and should be based on practical evaluations of actual performance and cost. Applied cost per unit of product, rather than per gallon or pound, is the only standard of cost that is meaningful.

It can be seen that an important factor in the design of an effective adhesive is the competence of the adhesive formulator company, which is responsible for assigning trained personnel, being alert to new technical developments, performing research, and promoting and maintaining high standards for its products and for the industry (4).

Selection of the Solvent Cement. There are three principal types of solvent cement for thermoplastics.

1. Simple Solvents. Simple solvents, or blends of solvents, are carefully selected for the specific plastic to be bonded. There must be adequate solvency to soften the plastic surfaces to such a depth that when pressure is applied, a slight flow occurs at every point in the softened area. The solvent must dry completely without bloom and without leaving behind high-boiling residues that will plasticize and weaken the plastic. Whereas low-boiling solvents give the fastest setting action and are generally relatively inexpensive, they are more likely to cause crazing (qv) of the plastic and otherwise reduce optical clarity in the bond. The rapid evaporation of the low-boiling solvents leaves the joint in a state of stress, and crazing, the formation of many tiny

Table 1. Solubility Parameters of Various Solvents and Polymers

Solvent	δ	Polymer	δ
		polytetrafluoroethylene (Teflon)	6.2
ethyl ether	7.4	dimethyl silicone	7.3
		polyethylene	7.9
		styrene–butadiene rubber	8.1
		polyisobutylene	8.1
		natural rubber	8.3
amylbenzene	8.5	polybutadiene	8.6
xylene	8.8		
toluene	8.9	polysulfide polymer (Thiokol RD)	9.0
		ester gum	9.0
ethyl acetate	9.1	polystyrene	9.1
methyl ethyl ketone	9.3	neoprene GN	9.2
trichloroethylene	9.3		
perchloroethylene	9.4	alkyd, medium oil length	9.4
		poly(vinyl acetate)	9.4
		nitrile rubber	9.4
		chlorinated rubber	9.4
		polysulfide polymer (Thiokol F and FA)	9.4
		poly(methyl methacrylate)	9.5
methyl acetate	9.6	poly(vinyl bromide)	9.6
		amino resin (Uformite MX61, Rohm & Haas Co.)	9.6
dichloromethane	9.7	poly(vinyl chloride)	9.7
1,2-dichlorethane	9.8		
dioxane	9.8		
cyclohexanone	9.9		
Cellosolve (2-ethoxyethanol)	9.9		
acetone	10.0	amino resin (Beetle 227-8 U-F, American Cyanamid Co.)	10.1
		poly(methyl 2-chloroacrylate)	10.1
		ethylcellulose	10.3
		poly(vinyl chloride-*co*-acetate)	10.4
		cellulose dinitrate	10.6
		poly(ethylene terephthalate)	10.7
		polymethacrylonitrile	10.7
		epoxy resin (Epon 1004, Shell Chemical Co.)	10.9
		cellulose diacetate	10.9
isopropyl alcohol	11.5	cellulose nitrate, ½ sec	11.5
		phenolic resin (BR 17620, Union Carbide Corp.)	11.5
dimethylformamide	12.1	poly(vinylidene chloride)	12.2
nitromethane	12.7	nylon-8	12.7
ethyl alcohol	12.7		
		nylon-6,6	13.6
		polyacrylonitrile	15.4
water	23.4		

cracks, is one way by which these stresses may relieve themselves, particularly in brittle, low-impact-strength plastics.

2. *Solvents Containing Polymer* (dope cements). These consist of bonding solvents that contain in solution a quantity of the same polymer that is being bonded. There is much less chance of crazing with a dope than with a simple solvent; the higher viscosity of the dope provides an advantage in handling; the solids deposited after

evaporation fill voids that may be present due to imperfect fit of the parts; and the bond strength is frequently higher than when pure solvents are used.

3. Solvents Containing Monomer (polymerizable cements). These consist of a reactive monomer, compatible and preferably identical with that of the polymer to be bonded, together with a suitable system of catalysts and promoters. They may be formulated as "dopes," that is, containing some polymer in solution. The mixture is compounded to polymerize either at room temperature or at a temperature below the softening point of the thermoplastic adherends. The monomer is usually a good solvent for its own polymer, and accordingly these cements provide solvent bonding as well as excellent specific adhesion after polymerization. They are particularly suitable for coarsely fitting parts, and, if they polymerize rapidly enough so that there is a minimum of evaporation, the resultant low shrinkage provides a fairly stress-free joint.

The selection of the best solvent to use for solvent cementing a given thermoplastic is facilitated by consideration of the solubility parameters of the materials (see also SOLUBILITY OF POLYMERS).

Solubility parameter, δ, is defined as the square root of the cohesive-energy density (qv), which is the amount of energy required to vaporize one cubic centimeter of the hypothesized liquid. Generally, a nonpolar molecule will require less energy (heat input) for evaporation and will consequently have a lower solubility parameter than the highly polar associated molecules. Each plastic material dissolves best in solvents whose solubility parameters are about equal to its own.

The solubility parameters of a number of solvents and polymers have been determined and reported in the literature (5,6). Some of these are listed in Table 1.

Employing solubility parameters, we can better understand why polystyrene ($\delta = 9.1$) is soluble in butanone ($\delta = 9.3$), but not in acetone ($\delta = 10.0$). Interesting illustrations are known; some plastics will dissolve in a mixture of two liquids, neither of which is itself a solvent. For example, cellulose nitrate (11) will dissolve in a mixture of alcohol ($\delta = 12.7$) and ether ($\delta = 7.4$), but not in either alone. This principle is also put to use in the design of solvent mixtures for cementing of plastics, which can be blended to a desired solubility-parameter range independently of the property limitations of individual solvents, which may not be available in both the desired solubility-parameter and drying-rate ranges.

For each of the specific thermoplastics to be discussed, a group of simple solvents, composite-blended solvents, or monomer-cement formulations will be suggested. These solvents have been found to have adequate solvency (sufficiently close solubility parameters) to produce workable "cushions" of softened plastic in a reasonable period of time. Since the evaporation rate of the solvent is also of importance in designing for minimum crazing and maximum speed of attaining an adequate bond, Table 2 provides the boiling points of all of the solvents and related materials that will be mentioned in the paragraphs referring to specific plastics.

Acrylic Plastics. Because of the optical clarity of cast acrylic sheeting, this plastic has been selected for the fabrication of articles of considerable size and complexity, such as the transparent plastic mock versions of mechanical devices, which permit "observation" of internal working parts (see ACRYLIC ESTER POLYMERS). These articles are made by cementing sections of stock parts, and, with care and practice, the transparency of the basic acrylic resin plastic can be retained throughout the joints.

Methylene chloride is a useful cementing solvent for acrylic plastics, producing joints of medium strength and requiring a soaking time of 3–10 min. With ethylene

Table 2. Boiling Points of Materials in Cementing Compositions

Solvent	Bp, °C
propylene oxide	34
methylene dichloride	39.8
acetone	56.2
methyl acetate	57.2
chloroform	61.7
methanol	64.8
carbon tetrachloride	76.5
ethyl acetate	76.7
ethanol	78.3
methyl ethyl ketone	79.6
benzene	80.1
ethylene dichloride	83.5
trichloroethylene	87.1
isopropyl acetate	88.7
methyl methacrylate	101
nitromethane	101.2
dioxane	101.3
toluene	110.6
nitroethane	114.0
butanol	117.7
acetic acid (glacial)	117.9
perchloroethylene	121.2
methyl Cellosolve	124.6
Cellosolve (2-ethoxyethanol)	135.1
ethyl benzene	136.2
xylene	138–144
methyl Cellosolve acetate (2-methoxyethyl acetate)	145.1
ethyl lactate	154.0
cyclohexanone	155.4
Cellosolve acetate (2-ethoxyethyl acetate)	156.4
diacetone alcohol (4-hydroxy-4-methyl-2-pentanone)	169.2
o-dichlorobenzene	180.4
p-diethylbenzene	183.7
butyl lactate	188.0
amylbenzene	202.1
isophorone	215.2
2-ethylnaphthalene	251

dichloride, the soaking time is 10–15 min. Since this is a slower-drying solvent, it requires a longer setup time and is less apt to produce cloudy joints. The shorter soaking times apply to plasticized acrylic sheets, and the longer soaking times to acrylic sheets of the unplasticized, heat-resistant type, such as those conforming to specifications for ASTM type II acrylic sheet.

Acetic acid cement consists of glacial acetic acid and requires a soaking time of about 1 hr at room temperature, or 2–5 min at 140°F. Although joints that are strong and have excellent optical properties are produced, the irritating fumes limit the use of this material, especially when hot acid is used, requiring vigorous ventilation and the use of protective rubber gloves.

The highly heat-resistant modified acrylic sheets are very resistant to solvents and crazing, and a polymerizable cement must be used. These adhesives are applicable to virtually all types of acrylic plastics, and comprise a mixture of solvent with catalyzed acrylic monomer. An example is a catalyzed 40:60 mixture by weight of methyl methacrylate monomer and methylene chloride. To each pint of the mixture is added 2.4 g of a 50:50 mixture of benzoyl peroxide and camphor (a stabilizer for the peroxide). After addition of the catalyst, the cement will have a useful life of up to 45 days if stored in a tightly closed container at not over 77°F. It is important that the monomer:solvent ratio be maintained after prolonged storage; this is usually done by measuring specific gravity and adding sufficient methylene chloride to replace that which has evaporated. The permissible range of specific gravity at 77°F is 1.16–1.20.

Additional cement formulations, and their adaptability to specific types of acrylic resin plastics, are detailed in the wealth of literature available from the resin manufacturers. This literature gives instructions regarding the possible necessity for annealing or heat treating the cemented joints, which, if done carefully, will remove or redistribute the last vestiges of residual solvent and thus increase joint strength. This must be accomplished at a low enough temperature to prevent the assembly from being warped, and the solvent from boiling and producing bubbles.

For *molded* acrylic articles that must be cemented, it is very important that the article be annealed prior to bonding to reduce residual stresses. Otherwise, the joint will be prone to unsightly crazing. Typical annealing cycles for thin sections are 2 hr at 140°F for "easy-flow acrylics," and 2 hr at 170°F for "hard-flow acrylics." Longer times will be required for heavy sections.

Cellulose Acetate. Acetate sheeting and molded articles can be readily cemented to pieces of the same plastic using the following solvents: acetone, methyl acetate, ethyl acetate, methyl ethyl ketone, dioxane, nitroethane, methyl Cellosolve, methyl Cellosolve acetate, ethyl lactate, Cellosolve acetate, and diacetone alcohol.

The above solvents may be used alone, or in combinations with each other in mixtures that provide a wider boiling range, or with addition of diluents including methanol, ethanol, and toluene. A commonly used composition is a mixture by weight of 70% acetone and 30% methyl Cellosolve.

Dope cements containing dissolved cellulose acetate are preferred when an imperfect fit of the parts requires filling. A typical formula is: 130 parts by weight cellulose acetate, 400 parts acetone, 150 parts methyl Cellosolve, and 50 parts methyl Cellosolve acetate.

A more general formula suitable for use with a wide variety of ingredients is as follows: 8–12 parts by weight cellulose acetate (low viscosity), 45–75 parts solvents with boiling point under 100°C, and 20–50 parts high-boiling solvents (over 100°C).

After being cemented, the pieces should be held under light pressure for 5–10 min, and the assembly allowed to stand at least 24 hr before subsequent operations, such as sanding, polishing, testing, and packing, are performed.

Cellulose Acetate-Butyrate and Propionate. These plastics may be cemented with any of the solvents listed for cellulose acetate above, and, in addition, methylene dichloride, chloroform, ethylene dichloride, isopropyl acetate, nitromethane, butyl acetate, Cellosolve, cyclohexanone, and butyl lactate.

The composition of dope cements is similar to the "general formula" given above for cellulose acetate with the exception that the plastic to be dissolved is cellulose propionate.

Cellulose Nitrate. Good results are obtainable in cementing this plastic with acetone as the solvent. Where it is desired that the highest degree of optical clarity be attained, medium-boiling ketones and esters should be employed.

Ethylcellulose. A series of solvent mixtures has been developed that provide strong bonds between pieces of ethylcellulose plastic. They may also be used with dissolved ethylcellulose, as dope cements.

Among the compositions are: toluene:ethanol, 80:20; benzene:methanol, 67:33; xylene:butanol, 80:20; ethylene dichloride (alone); ethyl acetate:ethanol, 60:40; butyl acetate:toluene:ethanol, 1:1:1; and toluene:methanol, 80:20.

Both of the surfaces to be joined should be wetted with solvent, and soaking time should be from 1 to 3 min prior to assembly. The surfaces are then brought into contact and held under light pressure, approx 5 psi, for about 10 min. The joints may then be air-dried or dried in an oven at a temperature of about 125°F.

Nylon. At room temperature, conventional solvents will not provide effective bonds on nylon. Such materials as formic acid or molten phenol may be used but are very difficult to handle. Phenol containing 12% of dissolved water may be used at room temperature; soaking time is about 15 min, followed by about 1 hr in clamps and a final set in the clamps for 5 min in boiling water. It has generally been found more suitable to bond nylon moldings with adhesives.

Polystyrene. Polystyrene is soluble in a wide variety of organic solvents, and complex assemblies of molded parts may be joined by solvents, dopes, and adhesives. However, polystyrene is a particularly brittle material and injection-molded parts are subject to crazing if bonded by low-boiling solvents.

Fast-drying solvents such as methylene chloride, carbon tetrachloride, ethyl acetate, benzene, methyl ethyl ketone, ethylene dichloride, and trichloroethylene cause rapid crazing of polystyrene, although the bonds are developed rapidly.

A suggested polystyrene cement for transparent joints consists of a mixture of about 25% of solvent boiling under 100°C; 40% of solvent boiling between 100 and 200°C; 30% of even higher boiling solvent; and 5% of dissolved polystyrene. Where airtight or watertight seals are required, or where void-filling properties are made essential by the joint design, the polystyrene content should be raised to about 15% by weight.

The "impact grades" of polystyrene are not as readily soluble as general-purpose polystyrene, but may be bonded using solvents with boiling ranges between those of ethylene dichloride and *p*-diethylbenzene in the following list of solvents for polystyrene: methylene chloride, carbon tetrachloride, ethyl acetate, benzene, methyl ethyl ketone, ethylene dichloride, trichloroethylene, toluene, perchloroethylene, ethylbenzene, xylol, *p*-diethylbenzene, amylbenzene, 2-ethylnaphthalene.

Cemented polystyrene releases solvent slowly, and at least a week, in many cases as much as a month, should elapse prior to putting the bonded article into service.

Poly(vinyl Alcohol). Both sheeting and tubing of this plastic can be cemented to itself with water, or solutions of about 15% of glycerin or a glycol in water, preferably at about 140°F.

Poly(vinyl Chloride) and Vinyl Chloride–Acetate Copolymers. These are a wide range of plastic materials that differ markedly in their properties. The homopolymer of poly(vinyl chloride) is difficultly soluble, and accordingly difficult to bond by solvent-cementing techniques. As the percentage of vinyl acetate in the copolymer is increased, solubility increases and the number of solvents available for cementing becomes quite large. The more highly plasticized compositions are also more readily bonded.

Where smooth rigid surfaces are to be joined, the preferred method of using solvent adhesives is to apply the cement to the edges of the two pieces while they are clamped closely together, and to permit the solvent to flow between them by capillary action. Ketones are the preferred solvents. Propylene oxide is usually included as an ingredient since it contributes to very rapid attack; however, it should be blended with higher-boiling ketones such as methyl ethyl ketone and isobutyl methyl ketone. It has been found also that addition of a moderate percentage of an aromatic hydrocarbon contributes to rapid softening action.

Vinylidene Chloride Copolymers. These plastics are quite inert chemically, and accordingly the choice of good solvents for cementing is greatly limited. Cyclohexanone, tetrahydrofuran, o-dichlorobenzene, and dioxane may be used. It should be noted, however, that these are among the relatively more toxic solvents employed in bonding plastics, and precautions regarding proper ventilation are very much in order.

A fairly long exposure time to the solvent is required, which may be determined by feeling the solvent-wet surfaces for the presence of tack.

Cementing Technique

A discussion follows of the bonding processes applicable to solvent cementing. These processes are applicable to all cementing solvents and are independent of the adherends. Since this is a special case, and for convenience, this section has been introduced at this point rather than included in the more fundamental and diversified discussion of adhesive bonding processes that follows later.

The surfaces to be cemented must be clean and prepared properly (employing mechanical and chemical treatments if needed). The surfaces should be fairly smooth and aligned as nearly perfectly as practicable, to avoid large voids. Where the problem of getting proper contact is aggravated by warpage, shrinkage, flash, ejector-pin marks, etc, a void-filling cement such as the polymerizable type should be employed (see Preparation of Adherend Surfaces, p. 516).

Adhesives should be applied evenly over the entire joint surface, preferably to both adherend surfaces, and care must be taken to prevent application to surfaces other than those to be joined in order to avoid disfigurement of the plastic parts.

The assembly should be made as soon as the surfaces have become tacky; this usually means within a few seconds after application. Enough pressure should be applied to hold the cemented joint until it has hardened to the extent that there is no movement when released. If the required clamping process actually deforms the parts, the stress due to springback of the flexible parts must be relieved before the joint hardens thoroughly, or else the joint may subsequently fail. This may be accomplished by releasing the clamps while the adhesive is just slightly "wet." Subsequent finishing operations must be postponed until the cement has hardened.

Care must be taken that the vapor from the solvent is not confined, since the vapor and its condensate may etch and craze other areas of the plastic parts. Adequate ventilation is further important in protecting personnel from toxicity hazards of the solvents employed, and particular care must be paid to the fire hazard accompanying the use of solvents.

Application Equipment. Cements may be applied by a variety of methods such as felt pad, brush, flowing equipment, dip, and spray gun. The dip and capillary (flow) methods are the most commonly employed.

In the *dip method*, one of the two parts to be joined is dipped into the cement in such a way that the solvent will act on the desired area for the optimum time. The parts should be assembled immediately after dipping and held under contact pressure. Excessive pressure should be avoided, or else the soft cushion of solvent-swollen plastic surface will be squeezed out at the edges of the joint, resulting in an unsightly assembly.

Each assembly requires a specifically designed dipping fixture and clamp. The dipping tray should be made of material that is inert to the cementing solvents. The areas of contact of cement can be controlled by lining the bottom of the cement tray with a felt pad, which is kept thoroughly wet with liquid solvent cement; or by placing a layer of glass-rod supports in the cement just below the level of the material.

Where the surfaces to be cemented fit very closely, the cement may be introduced into the edges of the joint by a brush, by a dropper, or by a hypodermic needle. The cement is then allowed to spread to the rest of the joint by *capillary* flow. Fine wires may be inserted into the joint when the parts are assembled in a bonding jig. After the cement is introduced into the joint and has reached all parts, the wires are removed. This procedure is useful also in removing air bubbles and in filling voids in joints made in other ways.

It is generally important to employ the minimum amount of solvent cement and the minimum amount of resulting "cushion" of softened plastic. The thickness of the cushion must be great enough to provide complete contact and the exclusion of air bubbles. An excess reduces the strength of the joint, unnecessarily prolongs the setting of the bond, and may result in an unsightly joint from squeezed-out material. Inadequacy of cushion and poor contact will result in bubbles and uncemented areas in the finished joint.

Preparation of Adherend Surfaces

In general, the surfaces to be bonded must be clean and dry, free of any loosely adhering films, and particularly free of substances that interfere with wetting such as oil, grease, mold lubricants, etc. For specific materials, the following discussion will be helpful. These are typical procedures, the use of which is suggested for critical exposure conditions. For general purposes, surfaces should be clean, dry, and free from grease, oil, and corrosion products. Plastic surfaces should also be free from mold-release and antistatic agents.

Metals. Careful surface treatment is absolutely essential in instances where the completed assembly will be subjected to high humidity, water immersion, or exposure to salt spray. A simple test of cleanliness is the observation that a water droplet makes a low or zero contact angle on the cleaned surface; alternatively, cold water may be poured over a suspected surface and allowed to drain off for at least 15 sec. There should be no break in the film of water nor any tendency of the film to crawl or pucker. After drying, bonding should be done immediately.

Aluminum must be degreased in trichloroethylene vapor; the air-dried degreased surfaces are immersed for 13–15 min at 145–165°F in a bath consisting of 30 parts by weight of water, 10 parts by weight sulfuric acid (sp gr 1.84), and 1 part by weight technical sodium dichromate. After a rinse in running tap water until the water is neutral, a final rinse in water at about 140°F is provided, and the surfaces are blown dry with warm air. (For removal of mill marks, an alkaline cleaner can be used prior to etching.)

Stainless steel is degreased in trichloroethylene vapor and immersed for 10 min at 160–190°F in a solution containing 3 oz sodium metasilicate, 1.5 oz tetrasodium pyrophosphate, 1.5 oz sodium hydroxide, 0.5 oz a surfactant such as Nacconol NR, and water to make 1 gal. It is then rinsed in fresh water and blown dry with warm air.

The above examples are selected from among alternative procedures. Many specific cleaning methods have been developed for various metals for various end-use assembly conditions. Refs. 7–9 are recommended for further study, as is reference to the technical literature of the adhesive manufacturer.

Wood. The surface of the wood must be clean and smooth, and the moisture content must be within the range that will be encountered in service, to minimize the development of stresses by shrinkage or expansion. Certain species of wood, particularly those containing appreciable amounts of oily or resinous extractives, may be improved in bonding ability by solvent or alkaline surface treatments.

Rubber. Vulcanized rubber that is to be bonded to rigid surfaces should first be cyclized on the surface. The cyclizing procedure works best with butadiene and substituted butadiene elastomers. Whether this is necessary in any specific case is best ascertained by a test bond. The following are popular and effective cyclizing methods:

1. The rubber is immersed in concentrated sulfuric acid (sp gr 1.84) for 5–10 min in the case of natural rubber, and 10–15 min in the case of synthetic rubber.

2. Alternatively, a paste of concentrated sulfuric acid and barites can be used. Sufficient barites should be added to the acid to give a consistency that will not run. Contact time is as above.

After having been washed thoroughly with water and dried, the brittle surface of the rubber should be broken by flexing so that a finely cracked surface is produced. It may be necessary for the rubber to be washed after cyclizing with dilute caustic solution (about 1.2%) to ensure neutralization of residual acid that, if not removed, will interfere with the curing mechanisms of many adhesives, weakening the bond strength.

Plastics. As a minimum, it is absolutely essential that the plastic adherend surfaces be clean and free of films of moisture, oil, release agent, antistatic coating, or other contaminants. Simple treatment with detergents, or solvents (so long as they are selected so as not to craze or haze the plastic surface), will generally leave the surface clean and smooth. With plastics, however, cleanliness alone will not assure optimum bond strengths.

The very smoothness of most plastic surfaces can be a detriment where high bond strengths are required. A smooth surface presents a minimum surface area, and since adhesive strength is a function of both the surface area in contact, and the chemical nature of the surface (polarity) in regard to both the adhesive and adherend, bond strengths may be increased by roughening techniques employing both chemical etching and mechanical abrasion, and by chemical treatments that change the surface composition. A variety of effective procedures have been developed and applied commercially.

For example, acetic acid has been used to etch nylon, strong alkalies to etch epoxy resins derived from novolacs, sodium to react with polytetrafluoroethylene (Teflon), and flame, chemical, and radiation treatments with polyethylene. In general, these methods provide excellent results and will be discussed subsequently in further detail. They do, however, have certain disadvantages in regard to the handling of hazardous

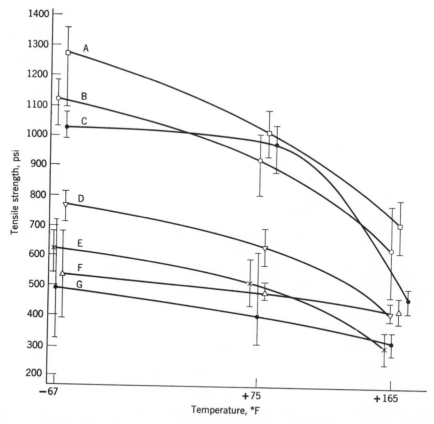

Fig. 3. Effect of plastic-surface preparations on bond strength. A. Vapor blasting. B Grooving and sandblasting. C. Raised knurling with vapor blasting. D. Raised knurling. E Grooving. F. Depressed knurling. G. Solvent wiping. Courtesy *Adhesives Age.*

chemicals, the requirement for highly skilled personnel, the practical limitations in applicability due to the size of the part or the dimensional tolerances, the hazard of contamination by residual chemicals, and in some cases high applied cost. These techniques are essential to improving the "specific-adhesion," or chemical, contribution to adhesive strength; in fact, with the inert crystalline plastics such as the fluorocarbons, adhesives with adequate specific adhesion to untreated plastic surfaces are extremely rare.

The chemical treatments are generally not applicable to laminates, which consist of alternate layers of reinforcing material and impregnating resin, since the chemical etchant may attack one of these materials preferentially. Laminates often carry a thin film of mold-release agent on their surfaces. This and the normal gloss must be removed by sanding or sandblasting. A subsequent blast of clean dry air to remove the dust is generally adequate with most plastic materials.

All plastics, indeed, will respond to mechanical abrasion treatments, which increase the surface area over which the adhesive acts. (For solvent cementing, of course, these treatments are not applicable.) If properly performed, mechanical abrasion may also produce tiny protrusions or hooks of plastic, which become mechanically embedded in the adhesive, further increasing the overall bond strength.

Some of the mechanical techniques that are applicable are hand sanding; machine sanding (much to be preferred); sandblasting, usually performed with a 40-mesh grit; vapor blasting (vapor honing, wet blasting), usually performed with 200-mesh or finer grit; hot or cold grooving, or knurling.

The effects of the various mechanical surface preparations on the strength of a nylon-reinforced epoxy laminate, which was bonded with a room-temperature curing epoxy adhesive in a double-step joint, are shown in Figure 3. It can be seen from this that techniques that provide a large number of small grooves are more effective than techniques providing a lesser number of the larger grooves. The curve for hand sanding was just about the same as that shown for sandblasting (10).

Chemical Treatments. The very inertness of the nonpolar plastics to most environments, including organic solvents, is accompanied by a resistance to adhesion. These plastics include the polyolefins, such as polyethylene and polypropylene, and the fluorocarbons, such as polytetrafluoroethylene (Teflon) and polychlorotrifluoroethylene (Kel-F). These polymers usually require additional chemical surface treatments; adhesives with adequate specific adhesion to the untreated surface of these materials are extremely rare. In general, "treated and bondable" grades are generally available on the market, and specific treatment processes are listed in the plastic manufacturers' technical literature.

With *polyolefins*, the successful treatments are believed to be oxidative in nature. They include immersion in a sulfuric acid–sodium dichromate solution, exposure to corona discharge, flame oxidation, immersion in aqueous solution of chlorine, or exposure to chlorine gas in the presence of ultraviolet light.

The chromic acid-oxidation method is probably the most convenient for use with molded polyolefin plastic parts of diverse contour. The treating solution has the following formulation (11): 75 parts by weight sodium dichromate, 120 parts water, and 1500 parts concentrated sulfuric acid. The sodium dichromate is first dissolved in water and the sulfuric acid is then added slowly to avoid overheating.

A typical immersion cycle with a freshly prepared solution involves immersion for 1 min at about 150°F. After removal from the bath, the treated plastic is first rinsed in fresh water, then in an alkaline detergent solution to neutralize the acid, and then rinsed again in water and dried. The treated surface should be air dried without wiping mechanically.

Using the above treatment on polypropylene bonded to aluminum with an epoxy–polyamide adhesive, tensile strengths of 760 psi were obtained with treated unabraded specimens. Without treatment, the bond strength was essentially zero. Without treatment, but with abrasion only, tensile strength was 393 psi, and surprisingly, the strength of bonds employing both abrasion and treatment was 695 psi.

For the *fluorocarbons*, reaction with metallic sodium dissolved in liquid ammonia (12) or in a solution of naphthalene in tetrahydrofuran (13) appears to give equivalent results and is suitable for the treatment of plastic parts of various contours. It has been speculated that this treatment reduces a very thin surface layer (on the order of 0.00004 in.) to elemental carbon, since fluoride ions may be detected in the sodium solution after treatment.

The composition of one of these solutions is as follows: 23 parts by weight metallic sodium, 128 parts naphthalene, and 1000 parts tetrahydrofuran. The naphthalene is first dissolved in the tetrahydrofuran; the metallic sodium is cut into slivers and added to the solution. Water and water vapor must be excluded. The Teflon or Kel-F part is immersed in the fresh treating solution for 1–5 min; a deep brown or black color indicates adequate treatment. The parts are then rinsed thoroughly in water, wiped clean with an acetone-dampened cloth, and dried.

In a bond-strength test of treated plastic bonded to aluminum, employing an epoxy–polyamide adhesive, typical results for Teflon were as follows: no treatment, no bond; treated, 1570 psi; abraded and treated, 1920 psi. For Kel-F, the results were: no treatment, 380 psi; no treatment, abraded, 1120 psi; treated, 2820 psi; abraded and treated, 3010 psi.

Many other treatment methods are described in the literature (14–24) and the reader is referred to these for further information. Among these treatments, the radiation, flame, and corona treatments are particularly suited for providing overall surface bondability of films that will require printing or laminating, although they have also been used to improve the printability of limited areas of molded plastic articles such as bottles.

Discrimination should be exercised in the selection of chemical treatments if, for example, polyethylene is to be used for food packaging.

Preparation of the Adhesive

It is of the utmost importance, in preparing a commercial adhesive for application, *to follow the manufacturer's recommendations.* Since these may provide information regarding storage conditions and safety, they should be read carefully by production personnel in the user's plant immediately upon receipt of the material. Safety instructions, which generally appear on labels as well as technical data sheets, should be clearly understood prior to opening the containers.

Many adhesives are inflammable; others contain ingredients that may be irritating, may have noxious odors, may cause skin sensitization, or may be toxic. The very versatility of adhesive formulation requires the use of virtually all of the chemical materials known to commerce, with their own specific attendant hazards.

Frequently adhesives are shipped in concentrated or dry form, requiring the addition of water, solvents, or other diluents to bring them to the optimum viscosity for application and proper wetting. Only those specific diluents and the specific proportions recommended by the manufacturer should be used. Overdilution may result in a weakened bond; underdilution may result in poor wetting; and improper selection of diluent may result in precipitation of active ingredients or in failure to attain proper drying conditions during the processing steps.

Where the adhesive is a two-component type, requiring the addition of a hardener or catalyst prior to application, the manufacturer's recommended mixing ratio should be strictly adhered to. The useful life of the catalyzed adhesive is known as the "pot life"; its limit is that time at which the adhesive, due to reaction with the catalyst, has increased in viscosity to the point where proper wetting of the adherends can no longer be obtained. The specified pot life must not be exceeded. In general, as the batch size mix is increased and the ambient temperature or the temperature of the ingredients is raised, the pot life is shortened; cooling, mixing in small batches, and mixing in shallow mixing vessels will extend the pot life. Many excellent machines are available for metering in the proper ratio, rapid mixing, and dispensing premeasured shots of short pot-life, highly reactive, multicomponent adhesive systems (Fig. 4).

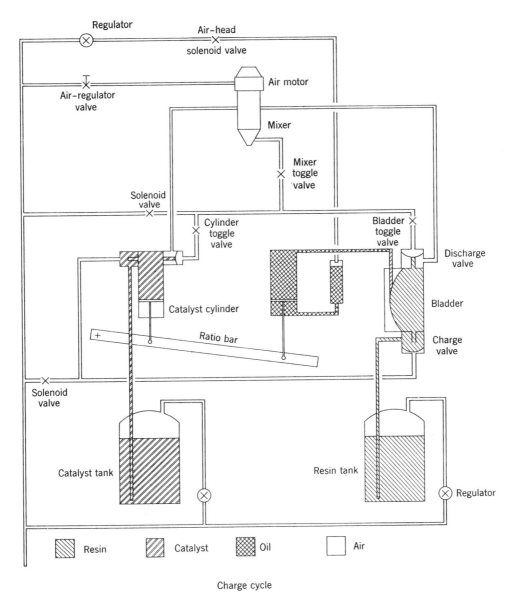

Charge cycle

Fig. 4. Metering mixer for multicomponent adhesives. Courtesy Automatic Process Controls, Inc.

For fluid adhesives, where the preparation involves simple addition of solvents or mixing in 55-gal-drum quantities, a simple propeller-type agitator, with an explosion-proof motor where solvents are involved, will be suitable.

For hot-melt adhesives, which must be melted to proper wetting viscosity prior to application, a simple electrically heated, thermostatically controlled melting pot is usually adequate. However, some degree of thermal degradation may result if the compounds are kept above the melt temperature for an extended period of time, and accordingly a number of devices have been developed recently that feed the hot-melt compound, in pellets or strips, into a very small accurately regulated melting

Fig. 5. Hot-melt adhesive applicator. Courtesy National Starch & Chemical Corporation.

chamber at exactly the rate required by the process, so that a minimum of material is above the melting point at any time. Figure 5 shows one such device.

With film adhesives and pressure-sensitive tapes, care should be taken to remove the protective release materials with which they are generally interleaved. The recommended procedure for "activating" the surface of the adhesive film to a tacky, fluid, wetting condition may involve wiping, brushing, or spraying with organic solvent, water, or liquid adhesive; in other cases this condition will be attained by the application of heat, usually within the bonding fixture.

The most important rule in handling adhesives prior to application is *cleanliness*. Scoops, weighing pans, mixing paddles, pumps, and pipe lines must be clean and free of any other adhesive, no matter how similar it may appear to be. Again, the manufacturer's instructions must be read and followed carefully.

Design of Adhesive Joints

Designing for adhesive bonding (25–27) is largely a matter of common sense. Two fundamentals should be observed: (*1*) The maximum amount of bonded area must be put to work. (*2*) Favorable geometry must be used for the joint design.

Maximum Area. The four basic types of stress (Fig. 6) encountered in structural bonding are: tensile, shear, cleavage, and peel. These stresses illustrate the use of the first fundamental.

In *tensile* loading, the forces are perpendicular to the plane of the joint and forces are thus uniformly distributed over the entire area. The entire joint is under stress at the same moment and all of the adhesive is at work at the same time. No portion of the joint is carrying more or less than its share of the load.

In *shear* loading, the stress is also distributed uniformly over the entire joint and all of the adhesive is put to work at the same time. In shear loading, however, the stress is parallel to the plane of the joint. This type of joint is most frequently used because it is more practical to accomplish.

In *cleavage* loading, not all of the adhesive is at work at the same time. As force is applied, one side of the joint is under great stress whereas the other is under no load at all. This type of joint cannot be as strong as a joint of comparable area under tensile or shear loading and therefore should be avoided (Fig. 7).

In *peel* loading, the stress is confined to a very fine line at the edge of the joint. In this case, even less of the adhesive contributes to strength than in the cleavage joint. Most of the adhesive is under no load and only a portion of the adhesive is at work. Even more than cleavage, this joint design should be avoided (Fig. 8).

Joint Geometry. Joint design is usually more complicated than indicated above; rarely, if ever, will a joint be subjected to only one stress. In practice, there is usually a

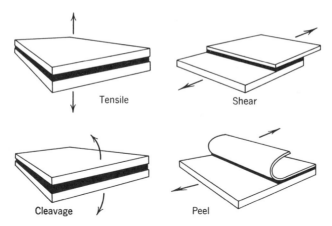

Fig. 6. Four basic types of stress encountered in structural bonding. Greatest strength is obtained when maximum area is used, as in tensile or shear loading. Courtesy *Materials in Design Engineering.*

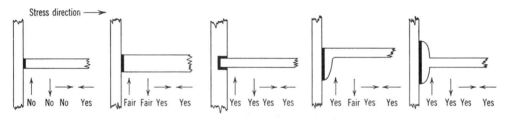

Fig. 7. Effectiveness of types of angle joints in resisting cleavage stresses. Courtesy *Machine Design.*

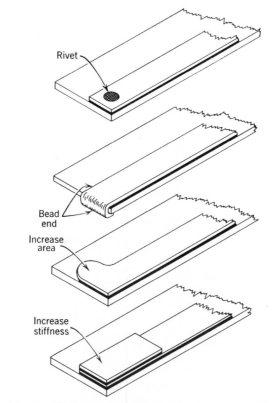

Fig. 8. Methods of avoiding peel stresses on rigid adhesives. Courtesy *Machine Design.*

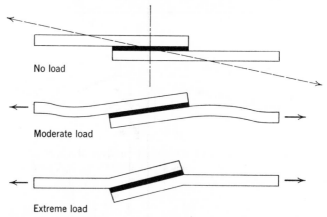

Fig. 9. Distortion caused by loading can introduce secondary stresses and must be considered in joint design. Courtesy *Materials in Design Engineering.*

combination of several different types of stresses. In some cases, distortion of adherends under load (Fig. 9) will introduce secondary stresses. In these cases the second fundamental, favorable joint geometry, becomes important. Following are the typical joints used in adhesive bonding.

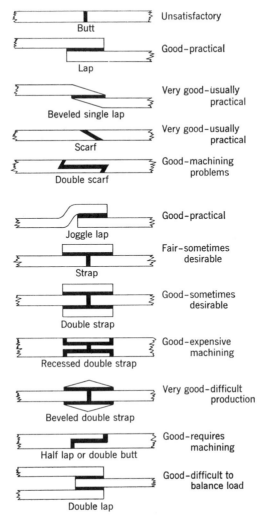

Fig 10. Lap joints. Courtesy *Machine Design*.

Lap Joints (Fig. 10). In adhesive-joint design, the lap (or shear) joint will be the most frequently encountered. Lap joints are quite practical and many designs are available for adhesive bonding, which gives a favorable stress distribution.

Because of the offset nature of simple lap joints, shear forces are not in line. Under moderate load, distortion of the joint occurs because the bond area will pivot normal to the load. At this point, an element of cleavage is introduced. When an extreme load is applied, such that fatigue occurs, definite bending of the material at the edges of the bond takes place. Then peel stresses are introduced. Thus, material distortion results in secondary stresses.

In these cases, the joint should be redesigned for maximum strength. There are three alternatives: (*1*) The joint can be redesigned to bring the load on the adherends in line. (*2*) The adherends can be made more rigid near the bond area to minimize cleavage. (*3*) The edges of the bond area of the adherends can be made more flexible for better conformance, thus minimizing peel.

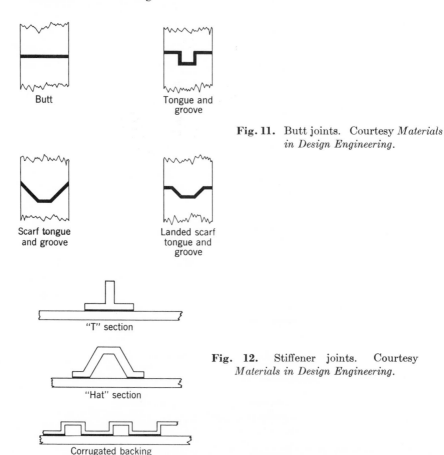

Fig. 11. Butt joints. Courtesy *Materials in Design Engineering.*

Fig. 12. Stiffener joints. Courtesy *Materials in Design Engineering.*

A beveled single-lap joint is more efficient than a straight lap joint. The beveled edge allows bending of the joint edge when distortion occurs under stress.

The half-lap, or double-butt lap, joint places the adhesive line in the same plane as the shear stress on the adherends. This type of joint, however, requires machining, which is not always feasible with thin-gage materials.

Double-scarf lap joints have better resistance to bending forces than double-butt lap joints. This type of joint, however, also presents machining problems.

The joggle lap joint seems to be the most practical. It places the adhesive line in the same plane as the shear stress on the adherends. In this type of joint, application of pressure for curing is easily accomplished and the joint can be formed by simple operations.

Figure 10 shows other types of lap joints that may also be used in bonding thin flat sections.

Butt Joints (Fig. 11). A straight butt joint is weak in cleavage. In the case where two flat, rigid rod ends are butt joined with an adhesive, any bending of the rods can, through leverage, exert tremendous cleavage forces on the joint. For this reason, recessed joints such as landed-scarf tongue and groove, conventional tongue and groove, and scarf tongue and groove are recommended. However, joints must have adequate clearance to facilitate machining and assembling operations. Thus,

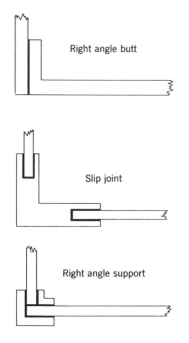

Fig. 13. Light corner joints. Courtesy *Materials in Design Engineering.*

adhesives with void-filling properties are required for butt-joint assembly, unless they are very carefully machined.

Landed-scarf tongue and groove joints have the advantage of acting as stops, which can be utilized to control the adhesive-line thickness. Scarf tongue and groove joints, on the other hand, make assembly easier in that self-aligning takes place when tongue and groove are joined.

Stiffener Joints (Fig. 12). Where large areas of thin-gage materials are used, problems of dents and deformation, "oil canning," waviness, and flutter are usually encountered. Stiffening of such surfaces is usually desired and can be efficiently and economically accomplished by adhesive bonding stiffening members to the large areas. When stiffening members are attached to thin metal sheets, as is common in aircraft construction, deflection of the sheets in service exerts peel stresses on the adhesives. If the flanges on the stiffening section can deflect with the sheet, minimum difficulty from peel is experienced. Either increasing the stiffness of the sheet or reducing the stiffness of the flange on the stiffening section will result in improvement.

Several types of stiffening members, such as T-sections, hat sections, and corrugated backing, are commonly used. T-sections are simple. Hat sections are more commonly used and have excellent rigidity. Corrugated backing results in the closest approach to flatness over the entire area.

Light Corner Joints (Fig. 13). By some joint redesign, it is possible to adhesive bond the corners of products made of light-gage steel or cored sandwich panels. The usual right-angle butt joint used for mechanical attaching is not applicable for adhesive bonding; it produces either cleavage or peel stresses. The use of supplementary corner reinforcement attachments, however, permits adhesive bonding and also seals the joint. Typical designs for adhesive bonding are slip and right-angle support joints. These joints also give the resultant structure an increased degree of rigidity.

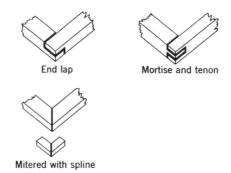

End lap

Mortise and tenon

Mitered with spline

Fig. 14.   Rigid corner joints.   Courtesy *Materials in Design Engineering*.

In a slip joint, use of a supplementary corner extrusion is best.   The extrusion can be as heavy as the desired rigidity demands.   The two edges of the part are bonded into the slip joint.   The use of interior and exterior right-angle supports is a variation from the slip joint.   However, this method requires indexing fixtures until adhesive is cured.

Adhesives in the form of films are not very suitable for these applications because of the complicated pressure fixtures required.   Adhesives with void-filling properties, such as epoxy types, are indicated.   The possibility of using a heat-curing epoxy adhesive depends on the heat resistance of the components being bonded.

Rigid Corner Joints (Fig. 14).   These are encountered when joining rigid members, eg, storm doors or decorative frames, and can also be adhesive bonded.   In these cases, butt joints would not be adequate for adhesives because of inherent racking or twisting stresses.   In order to make available the bond area required to resist such stresses, woodworking adhesive joint design, such as end lap, mortise and tenon, and mitered joint with spline, can be utilized.

End lap joints are simple but require machining.   Adhesives requiring pressure can be utilized for this application.   Mortise and tenon is an excellent type of joint but also requires machining.   A mitered joint with spline should be considered if the members are hollow extrusions.   The spline is usually die cast to loosely fit the interior of the extrusion.   In this case, a void-filling adhesive is also indicated.   See also Theory of Adhesive Joints, above.

Assembly

The final step of the bonding operation, assembly, consists of positioning the adherends and supporting them while the joint gains in cohesive strength.   This is done with the aid of bonding tools and fixtures.

In the broadest sense, bonding tools include devices used for preparation and cleaning of the surfaces of the adherends; fluidizing, mixing, metering, and dispensing the adhesive; applying the adhesive to the adherends; positioning and assembling the parts; providing temporary support and stability to the assembly while the adhesive gains cohesive strength; and maintaining those physical conditions, such as temperature, pressure, ventilation, artificial chemical atmosphere, etc, necessary to effect the chemical or physical changes that are taking place in the adhesive.

As a practical matter, bonding tools and fixtures may, in the case of a thermo-plastic hot-melt adhesive of the poly(vinyl acetate)–polyethylene–wax-resin type, be called upon to provide the heat of fusion, to apply the adhesive to the adherends, to maintain contact between the adherends and the adhesive, and to otherwise prevent separation. While the adhesive cools, its viscosity increases, and it develops sufficient cohesive strength to be able to support the assembly without outside assistance. In the case of a thermosetting hot-melt adhesive, such as an epoxy resin–dicyandia-mide curing agent powder adhesive, the bonding fixtures will have to perform the additional step of maintaining contact pressure and support for the bonded assembly over a sufficient period of time and at a sufficiently high temperature to effect a chemical reaction within the adhesive. In the case of a thermosetting poly(vinyl butyral)–phenolic film adhesive, for example, all of these previously noted steps take place, including a chemical curing reaction, with the additional requirement that the pressure maintained be substantially greater than contact pressure, since otherwise the gases (water and CO_2) liberated during the chemical reaction might rupture the bond.

The simpler steps involved in the use of aqueous or solvent-dispersed adhesives should not be ignored. Here the initial viscosity is inherently low. The function of the tools is to hold the parts in position and together while the viscosity of the ad-hesive increases owing to simple evaporation of the carrier fluid, or in special cases, to the absorption of part of the carrier fluid by a porous adherend.

The following are some applications of bonding processes and the tools and fixtures required in their execution.

In a simple laminating operation, one of the films is unwound from a stand and is coated with a thin fluid adhesive by a reverse roll coater. The coated film then passes through a drying oven that removes all or most of the carrier solvent. As it comes out of the oven, the adhesive film is hot and tacky. The material to be laminated with it is then fed over the hot adhesive film and the bonding is effected through a pair of heated nip rollers. The laminate then passes over a pair of cooling rollers to a rewind stand. In cases where both films, and particularly the second film, are nonporous, it is important that the drying conditions be such that the web leaves the oven vir-tually solvent-free. Here the temperature control on the nip rollers is critical, since they must maintain the adhesive in a molten fluid condition long enough to permit effective wetting of the second adherend. Where the second adherend is porous, such as with paper or cloth, a small amount of solvent may be left in the adhesive; the nip-roll temperature need not be controlled as critically, and the residual solvent will evaporate eventually through the porous material. Assemblies of this type have been used in the lamination of polyethylene to cellophane or to poly(ethylene terephthalate) film (Mylar), for food-packaging laminates; in this case, the more heat-sensitive polyethylene is the second adherend and it does not have to pass through the oven.

Many devices other than a reverse roll coater may be used for applying a fluid adhesive. Typically, these may be spray guns, and in many instances the adhesive will be preheated in the spray pot with an immersion heater in order to reduce its viscosity, thereby improving not only sprayability but also wetting of the sub-strate. Other types of coating heads are the knife-over-roll coater, engraved-roll coater, curtain coater, air-knife coater, etc.

The knife coater, although generally less accurate in its applied film thickness than a reverse roll coater, can use very high-viscosity coating doughs that are formulated to give a proper balance between adequate wetting of the textile surface and controlled mechanical penetration of the porous fabric. The coating is knifed onto one of the fabrics; the other material (which may be a different fabric or paper) is then joined through a pair of rollers to the wet adhesive-coated surface; and the laminate is then wound against a series of can dryers, that is, steel drums heated internally with steam. Tension is maintained on the web to assure positive contact and positive wrap against the cans. The more porous of the two webs is maintained on the outside, at the first can. This permits the bulk of the solvent to dry through this porous material. The diameter of the can and the speed of travel must be such that most of the solvent has been evaporated prior to the web's meeting the second can; otherwise, with a nonporous web on the outside and the metal can surface contacting the porous web, the evaporating solvent would produce blisters between the adherends. The final can is usually a cooling can.

In the relatively recent technique of "extrusion lamination," a hot-melt, solvent-free adhesive is melted in an extruder and cast in a thin film on one of the adherends. Before the adhesive sets, the second adherend is joined to the molten tacky adhesive surface through a pair of heated nip rollers. The laminate proceeds under tension over a pair of cooling rollers that set the adhesive, and is then wound up.

The use of hot-melt compounds has led to many interesting new forms of adhesives and to new dispensers to handle them. Thus adhesives are available as lumps, granules looking like molding resin granules, ribbons, rods, tapes, etc. They are fed to a heating chamber that melts them, maintains a small quantity in a molten condition, feeds the liquid adhesives to an applicator roll, applies the adhesive to one of the adherends, combines the other adherend, and delivers a finished laminate. Figure 11 shows one such dispenser.

In producing laminates on a production scale, with contact-bond adhesives, the two materials to be bonded are placed on a conveyor in alternate sheets and are fed under a coating head. (Until recently, this would typically be a reciprocating spray-gun device. The curtain coater has been replacing the spray gun for these applications. Here the adhesive is extruded under pressure in a thin film, vertically downward, and the sheet to be coated passing through this web picks up an accurately metered coat of adhesives. The excess adhesive falls into a reservoir, and is pumped back into the pressure pot after proper adjustment of viscosity and percent solids.) The coated parts, after leaving the coater head, are then again separated into two parallel lines and pass through ovens where the solvent is evaporated. These ovens may be of the infrared or circulating-air type. The solvent is dried out completely; retained solvent will weaken the assembly. The dried coated parts may then be assembled either hot or cold, depending upon the adhesive formulation. With contact cements, the immediate bond, adhesive surface to adhesive surface, is strong and develops very rapidly. The assembly is passed through a pair of nip rollers to assure good contact. Certain adhesive formulations, generally characterized as giving superior heat resistance in the bond, require that the assembly take place at elevated temperatures, and may require the use of additional heat-activating devices above the surface of the adhesive just prior to making the assembly.

Figure 15 shows the use of a vacuum bag, which surrounds the assembly and, when evacuated, assures that an even 14–15 oz/in.^{2} pressure is applied. This is particularly suitable for providing contact pressure to complex curved parts. One application of this technique, in the air-frame industry, produces contoured sandwich

Fig. 15. Bonding fixture in autoclave. Courtesy North American Aviation, Inc.

constructions with honeycomb core of aluminum or glass-fiber-reinforced resin and with aluminum skins. The adhesive is a fluid paste compounded from catalyzed liquid epoxy resins supported on a glass-cloth carrier. The vacuum bag in this case is made of poly(vinyl alcohol) film, although silicone rubber-impregnated glass cloth has also been used. After evacuation, the assembly may be heated in a circulating-air oven or in an autoclave, to heat cure the epoxy resin adhesive. Simple C-clamps are used to hold the assembly in position prior to evacuation.

A variation of the vacuum-bag technique is the use of a pressure bag. Such an assembly may be set up within a pressure vessel, which is subsequently closed and clamped exteriorly. Steam is introduced into the pressure bag, which then expands, applying pressure uniformly over whatever contour the assembly may have and simultaneously providing a source of heat. See also BAG MOLDING.

In discussing bonding fixtures, we cannot ignore the simple hydraulic press with steam or electrically heated platens. For a flat contoured laminate, this is an ideal and versatile device that applies controlled pressure and controlled temperature uniformly over the assembly. When the adhesive or laminating resins release fairly large volumes of gas during the chemical curing reaction, it is possible to open the press for a short period of time to bleed off these gases, closing it promptly to continue the cure. A hydraulic press can be used, for example, in the preparation of copper-faced plastic laminates for printed circuit baseboards. Typically, the copper foil is 1.4 or 2.8 mils thick, and is applied on one or both sides of the laminate. The adhesive may previously have been roller-coated onto the copper, and thoroughly dried or

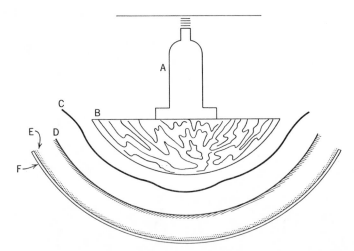

Fig. 16. Bonding jig for lining tanks with Kel-F fluorocarbon. A. Jack. B. Jig. C. Rubber. D. Laminate. E. Adhesive. F. Substrate. Courtesy Minnesota Mining & Manufacturing Company.

even partially cured. Alternatively, the adhesive, supported on a glass cloth or paper carrier, may be the first sheet of the laminate pad in contact with the copper. Typically, a number of sheets of resin-impregnated paper or glass cloth are assembled, with the copper on the outside, charged into the press, and cured. Depending upon the thermal and electrical characteristics required of the finished laminate, phenolic, epoxy, or other resins may be used as the saturants.

Special fixtures are used in producing bonded brakeshoes and bands for automatic transmission for automobiles. For the brakeshoes, the metal shoes, a sheet of dry-film adhesive, and the brake lining are assembled and a metal spring band is placed around them and tightened with a simple screw clamp. The assembly is placed in a circulating air oven that first melts the adhesive, wetting both surfaces, and then provides sufficient heat to effect a cure. The transmission bands are assembled by placing the spring steel outer-facing part, a sheet of film adhesive, and the friction lining around a drum, clamping them in place tightly, and then immersing them in a hot-oil bath, which activates the adhesive and provides the necessary curing temperature. Advantages of the use of adhesives include greater working life, because of the absence of rivet heads, and lower production costs.

Figure 16 illustrates a simple mechanically activated bonding jig, employed in bonding Kel-F fluorocarbon plastic linings to steel vessels with a liquid epoxy adhesive.

Whether heat is applied with a simple hand iron, by infrared or dielectric heating (qv), or a thermostatically controlled multi-opening platen press, and whether contact pressure is provided by a paper clip or a poly(vinyl alcohol) vacuum bag, the principle is the same; these are all bonding tools that assist in maintaining control of position and physical conditions during the setting of an adhesive.

Lamination of Plastic Films

A substantial volume of plastic material in the form of films is consumed in the production of laminates, for packaging as well as for industrial uses. The combination of the properties of several plastics, or a plastic and another material such as paper or foil into a single construction, provides properties not obtainable with any one film alone. This may be accomplished by lamination or by coating of one plastic on another.

The composite structure may provide improvements in strength, elongation, flexibility, scuff resistance, chemical resistance, resistance to transmission of water vapor, resistance to transmission of organic vapors or gases, or glossy appearance.

Lamination may also be used for building up the thickness of a homogeneous plastic. A laminate of two plastic films is actually a three-layer lamination, since the adhesive contributes to the strength and the resistance to vapor transmission, and also plugs up pinholes.

Laminating Adhesives. Solvent-dispersed adhesives, referred to colloquially in the converting trade as "lacquers," are available in a wide range of specific adhesions and compositions and are fast-drying. They are the materials of choice where both adherends are nonporous.

Emulsions and latexes are preferred, however, where one of the adherends is a moisture-permeable substrate. These have a high solids content, are free of fire hazard, and use inexpensive water as the diluent; also, the equipment used may be readily cleaned.

Hot-melt adhesives permit very fast production rates and can be applied with simple equipment. They contain no solvent or water to be removed. Formulation of hot melts is a technological field that is currently extremely active; new compositions, based on such ingredients as poly(vinyl acetate), polyethylene, and polyamides, modified with other resins to improve specific adhesion, are being made available in increasing number and complexity and are largely replacing the more familiar wax-based compositions.

For further information regarding adhesives see Adhesive Compositions.

Application of Laminating Adhesives. Since we are dealing in general with smooth surfaces on both adherends, the amount of adhesive applied is low, usually from 2 to 4 lb/3000 ft². Higher amounts, on the order of 4 to 8 lb/ream will be required with irregular porous stocks such as paperboard or cloth.

The equipment commonly used to apply laminating adhesives includes dip-roll, reverse-roll, gravure-roll, knife-coating, knife-over-roll coating, etc. See COATING TECHNIQUES.

Extrusion Laminating. This is a variation of hot-melt laminating, in which the hot-melt adhesive employed is a meltable thermoplastic polymer, such as polyethylene. In extrusion coating, the polyethylene coated onto the surface of the other adherend and cooled forms one of the films of the lamination. In extrusion laminating, the polyethylene is introduced in molten form between the two materials to be laminated and performs as the adhesive.

Just as it is necessary in making polyethylene "bondable" to perform an oxidizing operation on the surface, it has been found that extrusion coating and lamination with polyethylene is improved under conditions where the polyethylene surface is oxidized during the bonding process (28).

With adherends such as paper, the polyethylene will bond adequately without special preparation. However, with many plastic films such as polyester (Mylar) or cellophane, particularly if coated with a polyvinylidene resin, a primer or adhesion promoter should be employed. A typical adhesion promoter composition for polyethylene to be applied to steel or plastic films consists of isopropyl stearyl titanate, 5 parts by weight, and hexane, 95 parts by weight. This is applied to the substrate cured for 1 min at 140°C, and the polyethylene is extruded over it (29). Many other proprietary adhesion promoters have been described (30).

Commercial Laminations. A characteristic composite lamination is one which starts with cellophane film, coated with either cellulose nitrate or poly(vinylidene chloride) (to reduce organic vapor-transmission rate), which is then combined with polyethylene either by extrusion coating of polyethylene upon cellophane or by lamination with an adhesive. Typically, the adhesive consists of butyl rubber and polyterpene resins dispersed in hexane. The polyethylene, usually 1.5–2 mils in thickness, imparts water vapor and chemical resistance and the ability to be heat sealed. The resultant films are used for vacuum packaging of luncheon meats, cheese, and pharmaceuticals. More recently, the cellophane in this type of packaging lamination has been replaced by polyester film.

"Metallic yarn" is produced by slitting a laminate consisting of aluminum foil sandwiched between two layers of polyester film. A dyed lacquer adhesive is used to introduce color to the thread. More recently, the aluminum has been replaced by laminating a film of metalized polyester film to a clear polyester film. The adhesives used are compounded from acrylonitrile-copolymer synthetic rubber, vinyl resins, and vinylidene resins.

Laminations of poly(vinylidene chloride) film with paper and foil are widely used as bottle-cap liners. Acrylonitrile-containing synthetic rubber latexes are frequently used as the adhesive.

Laminations of thin ($\frac{1}{8}$ in.) sheets of polyurethan foam and various fabrics have become extremely popular as thermal insulating materials in garment fabrication. Crosslinking acrylic resin compositions are employed as the adhesive, and impart to the laminate the ability to be washed and drycleaned.

An infinite number of combinations is possible, and the number of laminates that have achieved commercial significance is extremely large. The above examples are illustrative only of a principle involved, namely, to take advantage of the best properties of two components in one finished structure with the aid of an adhesive.

Welding of Thermoplastics

The technique of uniting parts by joining their heated surfaces and allowing them to flow together, thermowelding (see also WELDING; HEAT SEALING; DIELECTRIC HEATING), provides an advantageous means of joining all thermoplastics, so long as they possess adequate heat stability.

A welded joint is usually a butt joint. The strongest and clearest welded joints are obtained with thermoplastics with the highest degree of polymerization. Residual monomer and low-molecular-weight polymers, as well as some plasticizers, may tend to volatilize during the welding with a resultant foamy and weak joint.

All of the common thermoplastics, when properly heat-stabilized, may be welded by many techniques. Exceptions are cellulose nitrate, which is thermally unstable, and certain fluorocarbon polymers that require special handling. For thin-film form, specialized equipment has been developed and used successfully for many years, to heat seal joints in producing such articles as envelopes and pouches for packaging use.

The major differences among the various methods of welding plastics lie in the method of applying heat to fuse the materials. The four principal welding methods are hot-gas, heated-tool, induction heating, and friction.

Hot-Gas Welding. This method has been used successfully with molded articles, particularly of poly(vinyl chloride), poly(vinylidene chloride), and polyethylene. It has also been used with acrylic plastic and polychlorotrifluoroethylene.

The pieces to be joined are placed in their proper relative positions and held firmly, with a gap of about $1/16$ in. between them. A welding rod of similar composition, which provides filler material, is directed into the joint with a slight steady pressure. A blast of heated air or gases from a welding gun, at about 400–600°F, is directed at the welding rod tip and the adjacent areas of the plastic to be bonded. The gas stream may be heated by either electricity or a gas flame, and the orifice temperature should be controlled by a thermoregulator. With a distance of 1–2 in. between the gun tip and the weld bed, a temperature drop of about 200°F will be encountered, which must be considered in establishing the orifice temperature for the gas as at least 200°F above the melting point of the plastic.

Since complete fusion of the plastics does not take place, as it does in metal welding, the welds are seldom as strong as the parent plastics. Typical weld strengths as a percentage of the ultimate tensile strength of the parent material range from 50% for high-density polyethylene, and 75–90% for poly(vinyl chloride), to 95–100% for low-density polyethylene.

Heated-Tool Welding. Both the heated-tool welding of sheet and molded articles, and heat sealing of films, make use of a heated tool to bring the plastic to fusion temperature. The edges to be united are brought into contact and allowed to cool under pressure.

Electric strip heaters, hot plates, soldering irons, and resistance blades are convenient means of providing high temperature locally.

In butt welding of sheets, an electrically heated blade is positioned accurately so that the two edges of the sheet are in contact with the blade. Accurate jigs and clamps are required to hold the sheets in alignment and provide sufficient bonding pressure, yet prevent buckling or flexing of the sheets. When the edges of the sheets have reached fusion temperature, the blade is raised and the sheets are pressed together and allowed to cool, forming the weld. This method has been used to butt weld poly(methyl methacrylate) sheets, for example, and provides a transparent strong bond.

A simple hot plate has been extensively used with poly(vinylidene chloride) plastics, for example, to weld sections of pipe and large tubing, and may also be employed in the assembly of molded articles of various thermoplastics in which dimensions do not need to be held closely. In this method, heat is applied to the surfaces to be joined by holding them in contact with the hot plate until a surface of molten material has been built up. As soon as the materials are sufficiently softened, they are removed from the plate and quickly joined and held firmly in their proper positions until the melted material has cooled to form a strong joint. A plate of solid nickel gives excellent results, because it will resist corrosion of the metal when adhering resin decomposes on its surface. Typically, a temperature of 275°C, 10 sec of contact, and pressure of about 8 psi will give satisfactory melting without excessive flash. The pressure between the two pieces during the actual welding union should be sufficient to press out air bubbles and to bring surfaces into intimate contact. The shortest possible interval should be maintained between removing the melted areas from the hot plate and joining them to form the weld.

The use of an electrically heated shoe attached to an ordinary soldering iron is one variation of this process, and is particularly suited for welding seams to produce continuous linings for tanks. The electrically heated shoe is passed slowly between the two surfaces to be joined and is followed by a hand roller that presses the molten surfaces together.

Heat Sealing of Plastic Films. This process is a variation of "heated-tool welding" in which the material is lapped as desired and heat is applied through the films, thus fusing the lapped material.

Equipment for heat sealing is of two general types: electrical-resistance elements that heat jaws, rollers, or metal bands, and high-frequency generators that make use of dielectric characteristics of the plastic itself to develop heat within the film. Specialized control equipment has been developed to control automatically the rate and area of heating, the pressure, and the cycle time, so as to provide strong tight joints without "burning through" of thin films.

Cellulose nitrate film is not heat-sealable because of its thermal degradation; cellophane and some polyester films are inherently not heat-sealable, but are available with coatings of heat-sealable polymers. Table 3 provides the heat-sealing temperature ranges for the common plastic films.

Table 3. Heat–Sealing Temperatures for Plastic Films

Film	Temperature, °F
coated cellophane	200–350
cellulose acetate	400–500
coated polyester	490
poly(chlorotrifluoroethylene)	415–450
polyethylene	250–375
polystyrene (oriented)	220–300
poly(vinyl alcohol)	300–400
poly(vinyl chloride) and copolymers (nonrigid)	200–400
poly(vinyl chloride) and copolymers (rigid)	260–400
poly(vinyl chloride)–nitrile rubber blend	220–350
poly(vinylidene chloride)	285
rubber hydrochloride	225–350
fluorinated ethylene–propylene copolymer	600–750

Tetrafluoroethylene polymer and chlorotrifluoroethylene polymer do not soften effectively below their temperatures of decomposition and cannot be directly welded satisfactorily. However, this can be overcome by the use of a flux, such as a mixture of the plastic with a fluorocarbon oil (31).

Spin or Friction Welding. This process utilizes frictional heat generated by rubbing two thermoplastic surfaces together to heat and fuse the areas. The strengths of the welds are essentially similar to those of the parent materials.

This process has the advantages of producing a weld of good appearance and strength; air (oxygen) is excluded from the joint since the heated surfaces are in direct contact with each other throughout the process; standard shop equipment, such as drill presses and lathes, may be used practically and economically.

Of course, the joint configuration is limited to circular parts, otherwise a circular weld area must be molded into and designed as a part of a noncircular part. In some cases, flashing may be required to ensure complete welds, and the flash may be unsightly. However, the joint may be designed in such a manner that excessive flash occurs only on an internal area of the article.

Spin welding involves rotating of one of the sections to be welded against the other, which is held stationary. Rubbing contact is maintained at a speed and pressure sufficient to generate frictional heat and melt the adjacent surfaces. When sufficient melt is obtained, pressure is increased to squeeze out all bubbles and to disperse the melt uniformly between the weld faces. Rubbing action is then halted to permit the weld to form, and pressure is maintained until the weld cools and sets. The frictional heat is of such intensity that it produces almost immediate surface melting without substantially affecting the temperature of the material immediately beneath the welding surface. A dynamic brake is required to assure that the rotational tool stops rapidly enough to prevent tearing or weakening of the weld as it cools.

Many practical relationships have been developed between the relative surface velocity, contact pressure, coefficient of friction, heat-transfer capacity of the plastic, and the joint design (32). The geometry of the joint is the most important factor in influencing weld quality. See also WELDING.

The diversity of the modern bonding tools, together with the sophisticated product of the formulator's laboratory, makes possible the exciting growth and variety of adhesive-bonded assemblies that are revolutionizing so many industrial designs. The adhesives themselves are siblings of plastics, and make possible the use of plastics as adherends in many articles and combinations where they would otherwise be excluded. Adhesive technology certainly merits the attention of every plastics engineer.

Bibliography

1. F. L. Darrow, *Story of an Ancient Art*, Perkins Glue Co., Lansdale, Pennsylvania, 1930.
2. I. Skeist, "Choosing Adhesives for Plastics," *Mod. Plastics* **33** (9), 121 (1956).
3. *ASTM Standard on Adhesives*, sponsored by Committee D 14, American Society for Testing Materials, Philadelphia, Pa.
4. *The Role of the Epoxy Resin Formulator in the Plastics Industry*, developed and prepared under the auspices of the Society of the Plastics Industry, Inc.
5. J. H. Hildebrand and R. L. Scott, *The Solubility of Non-Electrolytes*, Reinhold Publishing Corp., New York, 1950.
6. H. Burrell, "Solubility Parameters," *Interchem. Rev.* **14**, 3, 31 (1955).
7. H. Bickford, "Treatment of Metal Surfaces for Adhesive Bonding," *WADC Technical Report 55-87*, Part IV, February 1958.
8. S. N. Muchnick, "Treatment of Metal Surfaces for Adhesive Bonding," *WADC Technical Report 55-87*, April 1956.
9. R. S. Shane, T. L. Eriksson, A. Korczak, and D. B. Conklin, "Surface Treatment of Metals for Adhesive Bonding," *WADC Technical Report 53-477*, December 1953.
10. L. C. Jackson, "Preparing Plastic Surfaces for Adhesive Bonding," *Adhesives Age* **4** (2), 30 (1961).
11. N. J. DeLollis and O. Montoya, "Surface Treatments for Difficult to Bond Plastics," *Adhesives Age* **6** (1), 32 (1963).
12. *Teflon Tetrafluoroethylene Resin, Treatment for Bonding*, Information Bulletin No. X-75, Polychemicals Dept., E. I. du Pont de Nemours & Co., Inc., Wilmington, Del.
13. Minnesota Mining and Manufacturing Co., Brit. Pat. 765,284 (Jan. 9, 1957).
14. E. R. Nelson, T. J. Kilduff, and A. A. Benderly, "Bonding of Teflon," *Ind. Eng. Chem.* **50**, 329–330 (1958).
15. H. W. Arnold, "Etchant Makes Teflon Adhesionable for Bonding and Potting Applications," *Adhesives Age* **4** (2), 34 (1961).
16. H. S. Stern, "Ionizing Radiation Used in Making Fluorocarbons Adhesionable," *Adhesives Age* **3** (4), 26 (1960).

17. W. H. Schrader and M. J. Bodnar, "Adhesive Bonding of Polyethylene," *Plastics Technol.* **3,** 988–990, 996 (1957).
18. J. C. von der Heide and H. L. Wilson, "Guide to Corona Film Treatment," *Mod. Plastics* **38,** (9), 205–206, 344 (1961).
19. J. J. Bikerman, "Making Polyethylene Adhesionable," *Adhesives Age* **2** (2), 23 (1959).
20. W. H. Kreidl, U.S. Pat. 632,921 (March 31, 1953).
21. M. F. Kritchever (to Traver Corp.), U.S. Pats. 648,097 (Aug. 11, 1953), 683,894 (July 20, 1954).
22. F. N. Rothacker (to Modern Plastic Machinery Corp.), U.S. Pat. 2,802,085 (Aug. 6, 1957).
23. K. Rossman, "Improvement of Bonding Properties of Polyethylene," *J. Polymer Sci.* **19,** 141 (1956).
24. M. J. Bodnar and W. J. Powers, "Adhesive Bonding of the Newer Plastics," *Plastics Technol.* **4** (8), 721 (1958).
25. J. C. Merrian, "Adhesive Bonding," *Mater. Design Eng. Manual* **162,** 113–128 (Sept. 1959).
26. "Adhesive Bonding of Metals," *Welding Handbook*, Chap. 49, American Welding Society, 1959.
27. G. W. Koehn, "Design Manual on Adhesives," *Machine Design* **26,** 144–174 (April 1954).
28. J. G. Rote, Jr., Standard Packaging Corp., Clifton, New Jersey, personal communication.
29. W. W. Riches, "Titanium Esters as Adhesion Promoters," *Gordon Research Conference Lecture*, E. I. du Pont de Nemours & Co., Inc., 1961.
30. S. Gold, "Adhesion Promoters for Polyethylene," *Paper, Film, and Foil Converter*, Oct. 1958.
31. P. J. Wayne and W. M. Bruner, "Heat Bonding of Teflon Tetrafluoroethylene Resin," *SPE (Soc. Plastics Engrs.) J.* **11** (10), 28 (1955).
32. R. M. Stemmler, "How to Spin-Weld Acetals," *Plastics Technol.* **9** (5), 42 (1963).

Jerome L. Been
Talon Adhesives Corporation

DIELECTRIC HEATING

Electric currents, radio waves, infrared rays, and light are familiar examples of different electromagnetic phenomena. When electromagnetic energy comes in contact with matter (solid, liquid, or gas), it is partly or completely converted to heat energy: for example, an electric current may heat the wire through which it flows; infrared radiation may be used to cook food or bake paint; and laser beams may melt holes in metals. Electromagnetic energy at radio-frequencies can be used efficiently to heat many materials, including some which conduct electric currents very poorly or not at all.

The latter are of the class of materials called dielectrics; the heating process is termed *dielectric heating*. More generally, a dielectric material may be defined as one in which it is possible to store electrical energy by the application of an electric field; the energy is recoverable when the field is removed (see also ELECTRICAL PROPERTIES). Dielectric materials are usually very poor heat conductors. To heat such substances throughout their volume is very difficult with processes that apply the heat to the surface only. Electromagnetic energy in the radio-frequency (RF) range, on the other hand, can act below the surface of a dielectric material and heat *all* parts of the volume simultaneously, with substantially greater speed and uniformity of heating than with conventional methods. Other advantages of dielectric heating are that it can be turned on and off instantaneously; it is efficient and thus does not throw off a great deal of wasted heat; it can be precisely and accurately controlled with reasonably simple devices; it can heat selected sections of a part, leaving the remaining material cool. Dielectric heating equipment is easy to operate, is basically long-lived, and requires little maintenance.

Dielectric heating was used as early as 1880 by a physician, Dr. W. J. Morton (1), but its significance was first reported by d'Arsonval in 1890 and Tesla in 1891. By 1900, it was in practical use by doctors (who later named it *diathermy*) for treating parts of a patient's body well below the skin surface, with highly beneficial effects. Substantial industrial use did not start until World War II. Techniques were needed and quickly developed for setting resin glues in wood products, for preheating thermosetting plastics for molding; for welding vinyl materials, and for welding glass pipe (see WELDING) (2). After the war, its use increased rapidly in many fields. Dielectric heating is employed when simple heating is required, as in water removal from wood products (3), textiles, and foam rubber; freeze drying of food (4,5); thawing of frozen foods (6); and softening plastic materials for forming. It provides heat for chemical reactions: preheating thermosetting compounds for molding (7); setting resins impregnated in paper products (8); curing vinyl and polyurethan foam; curing resin glue in wood and paper products; and starting exothermic reactions in thermosetting resins being extruded continuously. It is also used in combination with mechanical processes for forming or welding plastics and plastic-impregnated or coated materials.

For dielectric heating, two ranges of radio-frequencies are used: For most processes, a frequency somewhere in the 1–200 megacycles per second (Mc/sec) range, usually called *high-frequency* or *radio-frequency heating;* for a small but increasing amount of work, frequencies above 890 Mc/sec, called *microwave heating.* The fundamental relationship for electromagnetic waves,

$$\text{frequency (Mc/sec)} \times 10^6 \times \text{wavelength (m)} = 3 \times 10^8 \text{ (velocity of light)} \quad (1)$$

indicates decreasing wavelength for increasing frequency. The wavelength for 30 Mc/sec is 10 m, commonly used for "high-frequency" heating. The wavelength for 1000 Mc/sec is 0.1 m, which is considered short for a radio wave, and is therefore called a "microwave."

In high-frequency heating, the material to be heated is usually placed between two electrodes. When high-frequency energy is applied to the electrodes, the material between the electrodes is heated fairly uniformly throughout its volume. In microwave heating, the energy is applied by *horns* or *waveguides*, and its effect decreases to a negligibly low value at some point below the surface, the depth of the penetration depending on the frequency and on the material being heated.

Theory

Dielectric heating, at any frequency, is the result of the interaction of electromagnetic energy with the various components in the atomic and molecular structure. An alternating electric field causes oscillatory displacements in the charged components of the dielectric, the energy for the motion being absorbed from the electric field. The charges carried by the oscillating components may be either permanent or induced. Each component resonates with the electrical field at a particular frequency that depends on its charge, mass, and structure. In gases and some liquids, this resonance phenomenon occurs at sharply defined frequencies, but in most solids it is spread over a broad range. All charged components undergo some oscillatory displacement at low frequencies, the motion becoming much greater at the resonance frequency or frequency range, and ceasing above it. In the ranges of resonance frequencies, considerably greater heating takes place than outside these regions.

In macroscopic terms, the dielectric material behaves in the following manner. When an alternating voltage is applied to a dielectric, a current (called a *displacement current*) flows through it, causing energy to be stored in the dielectric. The amount of current flowing, and the amount of energy stored, depend on the voltage, the frequency, the electrode configuration, and the chemical and physical structures of the dielectric material. The chemical structure determines to a large extent the *dielectric constant*, ϵ', of the material, a property defined as the ratio of the capacitance of a material in a given electrode configuration to the capacitance of the same electrode configuration with a vacuum as the dielectric. Its value for any material decreases with increasing frequency, showing decreasing response to the electric field.

In the case of a perfect, or lossless, dielectric, the displacement current leads the voltage by a temporal *phase angle*, θ, of $90°$. For an imperfect, or "lossy," dielectric, the phase angle is less than $90°$. The angle by which it differs from $90°$ is called δ, the *loss angle;* its tangent, $\tan \delta$, called the *loss tangent*, or *dissipation factor*, indicates directly the fraction of the stored energy which is converted into heat by the dielectric. The cosine of the phase angle θ is known as the *power factor*, and its value is approximately the same as that of the loss tangent for small loss angles, such as are characteristic of the usual materials heated dielectrically.

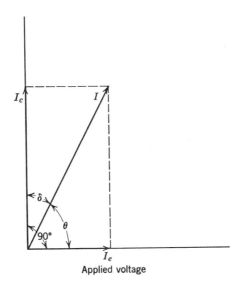

Applied voltage

Fig. 1. Representation of current components in an imperfect dielectric. I represents displacement current; I_c, charging current; I_e, effective heating current; δ, loss angle; θ, phase angle.

For calculations involving dielectrics, the displacement current can be resolved into two components, the *charging current* I_c, and the *effective heating current* I_e. I_c leads the applied voltage by 90°, and I_e is in phase with the applied voltage (Fig. 1). The ratio I_e/I_c represents the loss tangent, or dissipation factor, tan δ. The charging and the effective heating currents may be visualized as flowing in the branches of a simple circuit consisting of a "perfect" capacitor in parallel with a pure resistor (Fig. 2).

In the simplest form of dielectric heating, the material to be heated is placed between two metal plates. A generator applies to the plates a high-frequency voltage that sets up an electric field in and around the material. The material absorbs energy at a rate given by equation 2 (9),

$$P = 0.555 f E^2 \epsilon' \tan \delta \times 10^{-6} \tag{2}$$

where P = heat generated in watts/cc (dielectric loss), f = frequency in Mc/sec, E = field strength in V/cm, ϵ' = dielectric constant, and tan δ = loss tangent. This formula shows that the heating effect is directly proportional to the frequency, directly proportional to the square of the applied voltage, and directly proportional to the dielectric constant and the loss tangent. In most applications, the dielectric constant and the loss tangent are fairly constant over the dielectric heating frequency range, at a fixed temperature. Therefore, a "best frequency" need not be sought for; the desired heating rate is obtained by selecting a frequency range and voltage for which it is practicable to build equipment and for which a suitable electrode system can be designed.

The maximum value of voltage that can be used is limited by the voltage breakdown characteristics of the material, by its surroundings, and by electrode construction. Breakdown may occur inside the material, or in the space between the electrodes outside the material, damaging or destroying the material and melting holes in the electrodes. The breakdown is usually an arc of electrons or ions which concentrates the power of the high-frequency generator into a path of very small cross section. Arcing problems are reduced by careful electrode design and construction: The

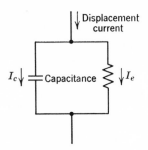

Fig. 2.   Simple circuit equivalent of a dielectric.

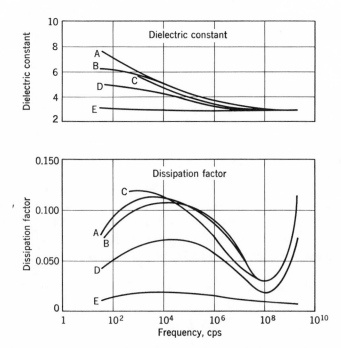

Fig. 3.   Effect of frequency of applied voltage on the electrical properties of commercial vinyl resins measured at room temperature: (A) Geon 2046, B. F. Goodrich Chemical Co.; (B) Koroseal 5CS-243, Goodrich; (C) Vinylite VG 5901, Union Carbide Corp.; (D) Saran B 115, Dow Chemical Co.; (E) Vinylite QYNA, Union Carbide (11).

electrodes should be designed so that as much as possible of the voltage applied to the electrodes is developed in the material being heated, and they should be made with rounded edges and corners wherever possible because sharp edges and points concentrate the voltage stresses and are the first places breakdown occurs.   It would appear that the frequency used should be as high as possible so that the lowest voltage can be employed, but there are limitations with this too.   At higher frequencies the generating equipment is more costly, and it is increasingly difficult to deliver the power from the generator to the material with good efficiency and control.   It also becomes increasingly difficult to maintain uniform voltage distribution over the entire mass of the material.

   The ease with which any material may be dielectrically heated is determined by

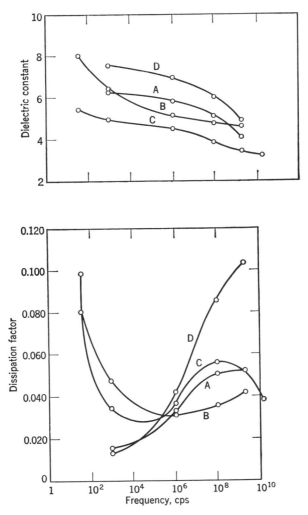

Fig. 4. Effect of voltage frequency on electrical properties of formaldehyde resins: (A) melamine–formaldehyde resin, wood-flour filled and plasticized; (B) melamine–formaldehyde resin with mineral filler; (C) cresol–formaldehyde resin with α-cellulose filler; (D) melamine–formaldehyde resin with α-cellulose filler (11).

its dielectric constant and its loss tangent. Values of these factors for several typical materials are given in Table 1 (10), for different frequencies. Notice that ϵ' falls with rising frequency, except for polytetrafluoroethylene; the sharper the drop, the higher will be the loss tangent. Coincident with the higher loss tangent shown for water at 3000 Mc/sec, there is a sharp decrease in the dielectric constant, not shown in the table. Polytetrafluoroethylene shows no change in ϵ', and the loss tangent is, as anticipated, exceptionally low. Changes in dielectric constant and dissipation factor over ranges of frequency and temperature are shown in Figures 3–5 for some other materials.

The product, $\epsilon' \times \tan \delta$, also called the *loss index* (or *loss factor*), ϵ'', shows most conveniently the combined effect of the two factors. Table 2 lists this product for many materials, and indicates the relative ease of heating. Loss indexes of 0.2 or

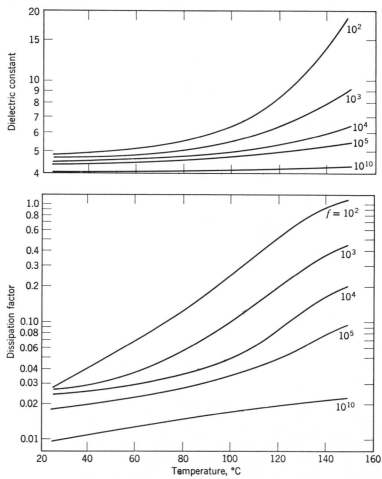

Fig. 5. Effect of temperature on dielectric constant and dissipation factor of phenol–formaldehyde resins (Resinox 10900, Monsanto) (Ref. 10, p. 409).

Table 1. Typical Dielectric Constants and Loss Tangents of Several Materials[a]

Material		Frequency, Mc/sec			
		1	10	100	3000
phenol–formaldehyde resin, 57°C	ϵ'	4.9	4.65	4.5	4.15
	tan δ	0.043	0.047	0.048	0.053
polyamide (nylon-6,6), 25°C	ϵ'	3.33	3.24	3.16	3.03
	tan δ	0.026	0.024	0.021	0.013
polyester, 25°C	ϵ'	3.11	3.04	2.98	2.85
	tan δ	0.0130	0.0160	0.0160	0.0100
polytetrafluoroethylene, 22°C	ϵ'	2.1	2.1	2.1	2.1
	tan δ	.... less than 0.0002			0.00015
ice, −12°C	ϵ'	4.15	3.7		3.2
	tan δ	0.12	0.018		0.0009
water, 25°C	ϵ'	78.2	78.2	78.0	76.7
	tan δ	0.040	0.0046	0.0050	0.1570

[a] Values derived from von Hippel.

Table 2. Relative Response of Various Materials to Dielectric Heating[a]

Material	Typical loss index	Good	Fair	Poor	None
ABS polymers	0.025		*	×	
acetal copolymer	0.025		*	×	
cellulose acetate	0.15	×			
diallyl phthalate polymer, glass-filled	0.04		*	×	
epoxy resins	0.12	×			
melamine–formaldehyde resin, cellulose filler	0.2	×			
phenol–formaldehyde resin, wood-flour filler	0.2	×			
polyamide	0.16		×		
polycarbonate	0.03		*	×	
polychlorotrifluoroethylene	0.025			×	
polyester	0.05		*	×	
polyethylene	0.0008				×
polyimide	0.013			×	
poly(methyl methacrylate)	0.09		×		
polypropylene	0.001				×
polystyrene	0.001				×
polytetrafluoroethylene (Teflon)	0.0004				×
polyurethan foam				×	
polyurethan–vinyl film		×			
poly(vinyl chloride), flexible, filled	0.4	×			
rubber, compounded	0.13		×		
rubber, hevea	0.015			×	
silicones	0.009			*	×
urea–formaldehyde resin	0.2	×			
water	0.4	×			

[a] Information derived from references 12 and 13 and from the dielectric heating experience of the author; (×) heatability in the 20 to 30 Mc/sec range; (*) response of the materials in the 70 to 100 Mc/sec range.

more result in good heatability; 0.08–0.2, fairly good heatability; 0.01–0.08, poor heatability; and under 0.01 there is little or no response. The lower the loss index the higher must be the voltage and frequency to obtain the required heating rate; for a material with a very low loss index, the required heating voltage at economically attainable frequencies is higher than the breakdown voltage.

Two calculations using equation 2 should demonstrate this. A typical application, to weld two square inches of two 0.005-in. thick films of poly(vinyl chloride) (flexible, filled), requires about 1000 W, which is 3050 W/cc. A reasonable voltage to use is 500 V across the double thickness of film, or 19,700 V/cm. By using a loss index of 0.4, from Table 2, the frequency is computed to be 35.4 Mc/sec. Equipment is readily available to deliver this voltage and frequency. For the same power input to a poly-tetrafluoroethylene load of the same size, the frequency required is 35,400 Mc/sec, because the loss index of polytetrafluoroethylene is 0.0004. Since equipment at this frequency is beyond the current state of the art, polytetrafluoroethylene may therefore not be dielectrically heated at present.

There are, however, techniques available for heating some materials with low loss indexes. Many of these materials show a rise in loss index with rising temperature.

With auxiliary heating means—radiant heat, hot air, or electrically heated platens—the temperature can be raised to the point at which the loss index is high enough to make the material susceptible to dielectric heating. A similar result may be achieved by placing a high-loss material in contact with the low-loss material; the high-loss material heats and transfers its heat to the low-loss load. Another approach is to mix additives or fillers with the low-loss material to raise the loss factor to a suitable level. Some examples of such additives are carbon black in rubber, sodium chloride in urea–formaldehyde glues for wood, and poly(vinyl chloride) in polyurethan foam—this last not only increases the susceptibility of the foam to dielectric heating but makes it more readily bondable to poly(vinyl chloride) sheet. Since the filler may affect the chemical, physical, or other properties of the base material, the type and the amount added are limited by the changes that can be tolerated.

Equipment

High-frequency and microwave heating equipment usually has five major sections: the power supply, the high-frequency generating system, the high-frequency transmission system, the control system, and the work applicator fixtures. The sections may be contained in one or more cabinets, depending on the application. The power supply section usually contains transformers, rectifiers, and switch gear, which convert the usual powerline energy (eg, 440 V, 3-phase, 60 cycles) to high-voltage dc energy, at 1,000–20,000 V. The high-voltage dc is fed to the high-frequency, or oscillator, section, which usually consists of a single high-power electronic tube (vacuum triode, tetrode, etc), with associated high-frequency circuits. In microwave equipment, the generating circuits are usually an integral part of the tube, called a magnetron; sometimes a klystron tube is used, with external circuits, or "cavities." The high-frequency voltages are actually generated in a capacitor–inductor combination. Energy is stored alternately in the capacitor and in the inductor—in the capacitor in an electric field, and in the inductor in the magnetic field of the current flowing in the inductor. Current flowing in the inductor charges the capacitor to one polarity; when it is fully charged the current stops flowing and then begins flowing in the opposite direction through the inductor to charge the capacitor fully in the opposite

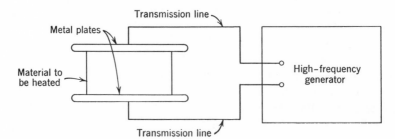

Fig. 6. Schematic representation of elements of a simple dielectric heating system.

direction. The vacuum tube acts as a switch between the power supply and this inductor–capacitor combination, switching current from the power supply at the appropriate times required by the capacitor–inductor combination.

The high-frequency voltage, up to tens of thousands of volts, is delivered through a transmission line to the work applicator fixture. These fixtures have many variations, of which the arrangement shown in Figure 6 is the simplest. Figure 7 is a picture of a small plastic preheater, which uses the arrangement of Figure 6, showing two plastic preforms on the lower electrode; the upper electrode is tilted backward for access. The upper electrode and hood are lowered and closed for the heating cycle. Figure 8 has similarly arranged plates, but opened up to give a gap above the material

Fig. 7. A 1-kW plastics preheater. Courtesy W. T. LaRose & Associates, Inc.

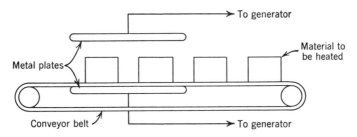

Fig. 8. Dielectric heating system with conveyor belt.

being heated; the power level can be adjusted by changing the height of this gap. It also shows a conveyor that can be used to carry materials between the plates on a continuous or intermittent basis. The conveyor belt must be made of a material that will not react with the load or be affected deleteriously by the electrode field or heat.   The belt might be a low-loss material such as silicone rubber, or glass fiber, or it might be a good conductor such as stainless steel.

The control section turns the power on and off; it may be used to set components in the other sections to adjust the power level to the requirements of the process. Meters are usually provided to indicate the electrical performance of the power tube. Protective relays are incorporated where needed to turn the machine off in the case of component failure or overheating, or other improper operation.   For process control, direct temperature measurements by conventional thermocouple techniques are not used because the thermocouple seriously distorts the heating pattern.   Instead, the heating rate is sensed indirectly, but quite accurately, by a dc meter indicating the power supply current, or by electrode voltage measurement, or both.   In batch processes the total heat input is controlled by setting the power controls and turning the power on for a fixed time interval.   Automatic controls are available to adjust the power level to a programed set of current or voltage values.

Although direct temperature measurement is not usually feasible, properties of the material (which change with the temperature) may be monitored for process control. For example, the dc resistance of the heating material can easily be measured; the resistance may vary widely with changing temperature, and so may be used to regulate the power level for temperature control.   The resistance level usually drops precipitately just before electrical breakdown (arc-over) occurs, and is very low after arc-over.

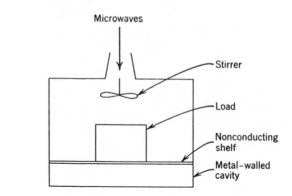

Fig. 9. Schematic representation of microwave heating system.

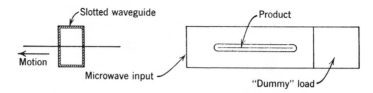

Fig. 10. Arrangement for dielectric heating with microwaves using waveguides.

Detection circuits are available to sense this change and to turn off the high-frequency power with such rapidity that the electrodes will be virtually unmarked (although the load itself may be damaged).   Other devices that measure the proximity of the work to an arcing condition can limit the occurrence of arcs.   This is a necessity where the electrodes are expensively machined members which must provide very precise mechanical pressures to the load.   Another method of protection against arc-over makes use of the "radio noise" generated by an arc.   A "noise" detector probe is placed near, but not necessarily in physical contact with, the electrodes and material; the probe circuits will turn off the power when arc-over occurs.

Although its power supply and control sections are similar to those of high-frequency generators, microwave equipment appears simpler because no high-frequency circuits are apparent.   The microwaves are generated within the magnetron tube structure, aided by an external magnet.   From the magnetron, power is fed to the applicator.   For bulk processing, a horn arrangement "sprays" the waves into a cavity that holds the load.   The cavity is an integral part of the electrical circuit and its design must take into account the frequency and load characteristics.   Energy not absorbed immediately by the load is reflected to the load by the metal walls, improving the heat distribution.   Figure 9 shows this arrangement schematically.   The non-conducting shelf above the bottom allows some energy to be reflected to the underside of the load.   The energy distribution in the cavity is not uniform and is referred to as a *standing field*.   A slowly rotating fan or "stirrer" also helps heat the load more evenly.   Microwaves can be transmitted through hollow round or rectangular pipes called "waveguides."   Products can be heated by placing them in the waveguides. Thin films can be run through the waveguides by slotting as in Figure 10.   The magnetron will burn out if the equipment is operated without a load; a safety device is therefore required—it is usually a "dummy" load (shown schematically in Figure 10) arranged to absorb some power when the normal load is absent.   By running a film through several waveguides in succession, reasonable efficiencies can be attained.   A waveguide can be terminated in a parabolic horn to spread the energy over a wider area, for heating thin film or bulk products (Fig. 11).

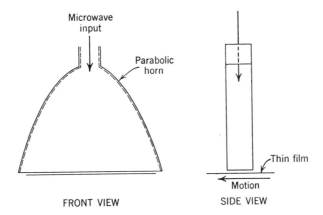

Fig. 11.   Parabolic waveguide for heating with microwaves.

For many applications, standardized equipment is available as complete processing packages.   Some applications are so specialized that equipment must be specially designed—the equipment manufacturer may build the entire machine, or he may furnish the generator, which the user will then connect to an electrode system of his own construction.   High-frequency generators are available in a wide range of output

Fig. 12. A 40-kW generator. *Upper half* is high-frequency generating section; at right, copper tubing inductor. *Lower half* is power supply; right half is power transformer and rectifier; at left, blower for forced-air cooling of power tube. Courtesy Faratron.

power ratings, from about 50 W up to many hundreds of kilowatts. Up to 2 or 3 kW can be obtained at up to 200 or 300 Mc/sec; 25 kW at 100 Mc/sec and 100 kW at 30–40 Mc/sec are also available, although the operating frequency is usually lower for these and for higher powered units. Microwave generators can be had at 900 or 2500 Mc/sec, with outputs up to 25 kW. For continuous-flow processing, generators and electrode systems can be set alongside each other to meet larger power requirements. However, it is extremely difficult to make more than one generator operate on one electrode system. Batch processes therefore require a single generator of large enough capacity to supply the entire heat requirement. Figure 12 shows the inside of a 40-kW generator.

Applications

Preheating. The simplest application of dielectric heating is for preheating materials for compression or transfer molding; preheating is not only desirable but necessary in some cases (see MOLDING). Preforms or powders of phenolic, urea, melamine, or alkyd resins, polyesters (including glass fiber-reinforced materials), allyl polymers, and others are heated to about 250 to 325°F, in 15 to 60 sec. The rapid, thorough heating reduces mold pressure requirements by at least 50% and cuts cycle times 15 to 85% below the values needed when no preheating is used. Substantially higher product quality frequently results; rejects are fewer and mold life is increased (7).

Preheaters operate in the 20–100 Mc/sec range, present practice leaning toward the top end of this range for greater heating speed, especially for lower-loss materials. They are available in many sizes, as manually loaded units (Fig. 7), or with mechanized material-loading equipment for automatic operation (Fig. 13). Power output requirements are figured at 1 kW to heat one-third to two-thirds of a pound of material to molding temperature in 1 min. Preheaters are also used for moisture removal from plastic and wood products; a typical large special-purpose equipment is shown in Figure 14. For such applications power requirements can be estimated closely by standard thermal calculations, taking into account such factors as specific heat, heat of vaporization, and heat losses. Use of a preheater for defrosting large rubber bales is shown in Figure 15.

Dielectric preheaters are now being used in casting (qv), potting, and curing of epoxy resins, polyesters, polyurethans, and vinyl plastisols (see VINYL DISPERSIONS). In a typical process, an epoxy resin–hardener system for casting is heated in about 30 sec from room temperature to just below its curing temperature, and then poured into a heated steel mold. The filled mold is then heated for 30 min in a conventional oven, which gels the casting sufficiently to permit removal from the mold for postcure.

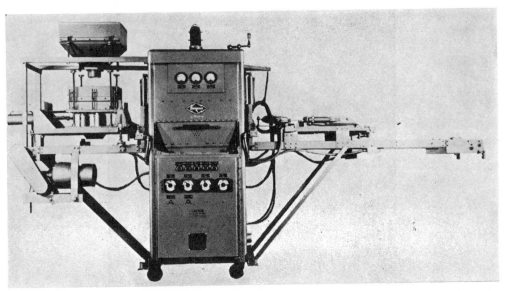

Fig. 13. Preheater equipped with hopper and feed mechanism and with 7.5-kW output for automatic heating of powdered materials. Courtesy W. T. LaRose & Associates, Inc.

Because the time the mold would remain in the oven without dielectric preheating is about 1.5 to 2 hr, preheating results in greatly increased productivity for the mold, which may be complicated and costly. Other benefits of this process are more uniform heating, greater freedom from air bubbles, fewer thermal degradation problems, and less restriction on mold design (14).

Foam Curing. An arrangement for curing of vinyl foam by dielectric heat is sketched in Figure 16. Unfoamed vinyl compound is poured into a 4 × 8 ft tray consisting of an aluminum base with silicone–glass fiber-laminate sides. (This is a rigid low-electrical-loss dielectric material, needed to contain the foam without dis-

Fig. 14. Unit for preheating stacks of fiberboard panels prior to pressing into hardboard. Equipment is rated at 200-kW output continuous duty. Electrode system, in compartment on right, is a low-pressure press with 5×10 ft platens. Heating cycle of about 2 min is required to dry to zero moisture and preheat to 325°F. Courtesy Votator Division, Chemetron Corporation.

Fig. 15. Dielectric heating equipment with 30-kW output being used for defrosting 250-lb rubber bales prior to processing. Courtesy W. T. LaRose & Associates, Inc.

torting the high-frequency field, and without itself being deleteriously affected by the field or the thermal conditions in the oven.) Several of the trays are stacked in an oven, each tray under its own electrode. Because the liquid vinyl resin is only a fraction of an inch thick, it cannot be heated efficiently with the electrodes spaced for the full-foamed thickness of about 6 or 8 in. However, it is thin enough to be heated rather uniformly by the gas-heated hot air that is circulated around the trays. The hot air provides sufficient heat for complete foaming of the liquid, but this heat is totally inadequate to cure the full thickness of the foam in any reasonable time without serious degradation of the outside of the slab. Dielectric heating provides the ideal solution to this problem. While the hot air circulation is maintained around the trays, the dielectric heater is turned on for about 10–20 min, giving an extremely uniform cure throughout the 8-in. slab. Dielectric heating could also be used in the foaming stage. The electrode would be placed close to the surface of the unfoamed material,

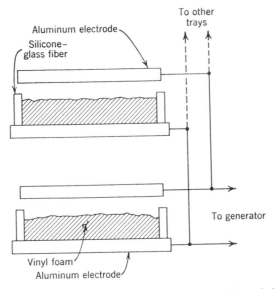

Fig. 16. Arrangement for curing vinyl foam by dielectric heat.

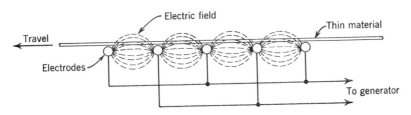

Fig. 17. Arrangement for drying thin material by dielectric heat.

and moved up as foaming progresses. The voltage might be varied simultaneously for quickest heating.

Drying. In drying applications, uniformity of dielectric heating is less critical. This is because water has a far higher dissipation factor than the usual material being dried. As the wet material is heated, the water evaporates and the rate of energy absorption in the dried areas drops drastically. The wet spots continue to absorb energy at a high rate. The material will usually dry quite thoroughly throughout its volume before the temperature of any of the drier parts rises considerably above the boiling point. For bulk drying, dielectric heating has great speed and uniformity, but it is also useful for thin materials. Figure 17 shows an arrangement for heating a thin web of material that has been in use since 1960 for drying photographic paper after development and rinsing. The electrodes consist of parallel rods of metal, such as stainless steel, which are connected to alternate terminals of the generator. The dashed lines indicate how the electric field is set up between the electrodes, causing what is generally called *stray-field heating*, because the material heated is not directly between the electrodes. The heating pattern is uneven; the web must be moved for best heat distribution. Although the surface closest to the electrodes dries first, the entire paper is dry in less than 10 sec. This system is now being tested in papermaking as a possible replacement for drum dryers. Experiments with a 30-kW unit have been so successful that two 150-kW generators are being constructed for operation in regular production. Dielectric drying produces higher quality paper in several respects: more even moisture distribution, better surface finish, and improved mechanical qualities. The dielectric dryer requires far less space than conventional equipment of equivalent production capacity (15,16). In place of the paper, a thin Teflon–glass fiber belt can be used to carry objects to be treated; the electric field acts through the belt with little diminution. This arrangement is being used to set poly(vinyl acetate) glue on the backs of books placed glued surface down on the belt.

Wood Gluing. Figure 18 shows another stray-field application, in which a panel of wood is glued to a wood block. The loss index of the resin is such that most of the heat is developed in the glue and very little in the wood, even though the glue line is somewhat removed from the electrodes and the voltage on the wood is slightly higher than on the glue. Mechanical pressure is applied through insulating members to attain a good bond. Figure 19 shows how glue is set in *edge bonding*. Wood sections several inches wide, from 0.5, to several inches thick, and several feet long, are squeezed between electrodes. Wood slabs 4 × 8 ft can be produced in a processing time of 30 to 60 sec; without dielectric heat it would take several hours. Power requirements cannot be figured easily because of the many variables involved, one of which is the

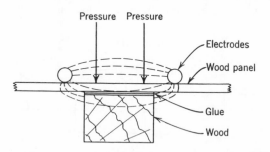

Fig. 18. Wood gluing by dielectric heating.

absorption of heat by the wood from the glue. It is generally estimated that 1 kW will set about 50 to 200 in.^{2} of glue line in 1 min.

In a process similar to edge bonding, kraft paper sheets are compressed and glued into a stack. When the glue is cured, the stack is expanded to make honeycomb core stock, to be used for later lamination to sheet materials to produce strong, rigid, lightweight panels. Figure 20 shows a 40-kW stack-gluing dielectric heater and press unit, and an expanded stack. A 200-lb stack, 10.8 × 18 × 24 in., can be glued up in

7 min on this machine; with conventional heating, a curing time of about 24 hr would be required.

Sealing. Dielectric heating frequently presents a problem of heat flow in reverse to that usually encountered. Although the load is heated uniformly, its exterior is cooler; its outside faces lose heat to the atmosphere and to the electrodes, or belt, or other mechanism that may be in contact with it. In some cases, this heat

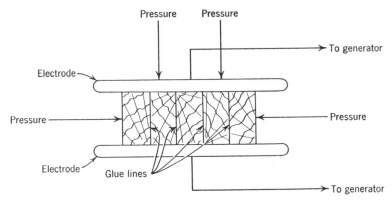

Fig. 19. Edge bonding of wood by dielectric heat.

Fig. 20. Paper honeycomb core gluer, expanded-stack hydraulic press, and 40-kW generator automatically squeeze and cure glues, and eject finished compressed honeycomb. Courtesy Faratron.

loss aids the process, and in others it must be compensated for.   The welding, or sealing, process shown in Figure 21 illustrates both effects.   Two pieces of 0.004-in. thick vinyl film are to be sealed together.   A metallic electrode, usually of brass, perhaps 1/16-in. thick and several feet long, squeezes the film against the steel bed-plate of a pneumatically operated press.   High-frequency voltage is applied, causing the plastic to heat and melt.   The brass and steel do not get hot under the influence of the high-frequency energy.   Since they are in intimate contact with the vinyl film

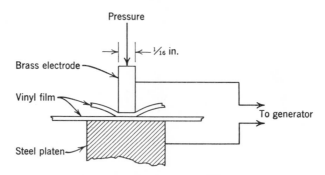

Fig. 21.   Arrangement for sealing film.

Fig. 22.   General-purpose dielectric film sealer, with pneumatically operated press mounted on a 10-kW, 30-Mc/sec generator.   Courtesy The Thermatron Company.

Fig. 23. Automatic dielectric sealer. A web of wadding is fed between two webs of vinyl film through a 75-ton press, which seals the three webs in a quilting pattern. A 35-kW, 30-Mc/sec generator (not shown) is used. Production rate is 24 seals/min. Courtesy The Thermatron Company.

and since the film is very thin, they absorb a good deal of the heat generated in the plastic. To compensate for the loss, several times more heat must be developed in the film than would be needed for melting it. However, after the plastic is melted and the power is turned off, the cool metal electrodes rapidly refreeze the plastic. Without refreezing under pressure, a poor bond would result. An additional advantage is that the outer surfaces of the film can be kept below the melting point, and so maintain their original characteristics. A manually loaded plastic film sealer is pictured in Figure 22, and an automatic machine in Figure 23. Total cycle time (heating and cooling) for these equipments can be a fraction of a second or several seconds, depending on the application.

Vinyl film and sheet are sealed by high-frequency heating to form a great variety of products, from wallets to swimming pools. Fabrics of cotton, nylon, glass fiber, or other materials, coated or impregnated with heat-sealable materials such as plasticized poly(vinyl chloride) or polyurethan, are sealed to form products requiring greater strength than can be obtained from the unsupported films. A striking example to see is the "air-house," an inflatable structure which typically can be constructed to provide a shelter 60 ft wide, 30 ft high, and hundreds of feet long, and is made almost entirely of dielectrically sealed poly(vinyl chloride)-coated nylon fabric; kept inflated with small air compressors and held in place with guy wires, it can easily support high wind and snow loads.

A large-scale application for dielectric heat sealing is in the manufacture of the inside door panels for automobiles. The panels are made of a supported or unsupported vinyl top layer, an interlayer of padding, and a bottom layer of paperboard about 0.1 in. thick. For good bonding, the padding may be impregnated with a vinyl or other adhesive, and the board is adhesive-coated.

Polyurethan and vinyl foams may be sealed between vinyl sheets for products such as pillows. Although vinyl does not "weld" to polyurethan, the wanted seal can be obtained because the vinyl sheets will weld to each other through the polyurethan foam if enough pressure is used and if the foam has a low density. As was previously mentioned, mixing poly(vinyl chloride) into the polyurethan will help the sealing process.

An unusual application of dielectric heating is the sealing of a cotton shirt label to a cotton shirt. The label has woven through it a polyester thread that, heated by high-frequency energy, is made to melt and flow into the interstices of the cotton fabric. The mechanical interlocking of the polyester with the cotton textile quite effectively bonds the label to the shirt. Similarly, a strip of nylon film has been used to seal two layers of the cotton fabric tape on zippers.

Many woven and nonwoven thermoplastic fabrics can be dielectrically sealed. Poly(vinylidene chloride) woven fabric has been used for many years in heat-sealed industrial products, as, for example, a fuel filter for automobiles. Dielectric heating equipment has recently been made available to seal some nylon and cotton–polyester textiles.

Microwave Heating. Microwave heating is most used in food processing, but there are some industrial applications and there has been much experimentation for new uses. *Domestic and commercial ovens* are employed for rapid food preparation—many of these contain conventional coils for radiant heaters to brown the food because the microwaves do not heat the outside enough to compensate for surface heat losses. Microwaves are also used for *food processing at the factory level* (6,17). Experiments have been made with *preform heating by microwaves*.

Moisture removal requires much energy, which can be cheaply supplied by conventional methods for most of the drying but removing the last small amount of moisture has usually been costly and space-consuming; microwaves do this efficiently and the equipment requires little space. Photographic film is now partially dried, after development and rinsing, by hot air and squeegee, then finish-dried by microwaves. In film manufacture, microwaves are being used to dry the buildup of emulsion on the edges.

There are other interesting uses: glue is set in the production of multisheet business forms; banks dry the magnetic ink printing on checks; glue is dried on paperboard for cartons; glue for "perfect" bindings (bookbacks glued without sewing) is set; and polyurethan foams are cured (18,19).

Economics

Because dielectric heating is qualitatively so different from all other methods, cost comparisons are usually meaningless. In the limited number of uses where similar results can be obtained by more conventional forms of heating apparatus, dielectric equipment is generally higher in cost. However, high-frequency units are more efficient as energy converters, putting into the load 50 to 70% of the power-line input, and, unlike ovens, no preheating of the equipment is needed (microwave equipment is far less efficient). Electric power costs will therefore be less than for other types of electric heating; the high efficiency can make it competitive with fossil-fuel heating in some cases. There are other cost factors to be considered: Plant space requirements are low, both for equipment and materials—other treatment methods may use more space-consuming equipment and may require space for materials to "cure" or "set" or "thaw" for hours or days. In some cases plant capacity can be multiplied without space increase by substitution of dielectric heaters for less efficient fossil-fuel ovens. The high efficiency allows the equipment to run cool; it is comfortable to work near and means of removing excess heat are not required. Equipment costs run from approximately $1000/kW for very small units and to about $250/kW for the highest powered machines. The electrodes and materials-handling devices may be inexpensive, or, for complicated processes, they may cost more than the generating equipment itself. Maintenance costs for dielectric heaters are minimal, usually much lower than for other heat sources. Although the dielectric heating machine has unfamiliar high-frequency circuitry, its control and power circuits are relatively unsophisticated; maintenance and trouble-shooting can usually be done by plant electricians or mechanics who have had a little instruction from the equipment manufacturer. Generator parts replacement costs for high-frequency units are around 1–2¢/hr/kW of output power, and about 2–3¢ for microwave equipments. Most electrode systems rarely need service or replacement.

Safety

The dielectric heater can radiate energy which might interfere with aircraft communications, television, or other radio services. To minimize interference, allowable radiation limits and methods of measurement are specified by the Federal Communications Commission's Rules and Regulations, Part 18, for Industrial, Scientific, and Medical Equipment. Compliance with these rules is required for operation of equipment in the United States. These rules permit unlimited radiation on certain very narrow frequency bands, or limited radiation on any other frequency. Equipment for operation on an assigned frequency band requires high-stability circuits for frequency control, and must be designed to prevent radiation of energy at all other frequencies generated internally because all dielectric heaters simultaneously generate their operating frequency and its harmonics. This type of equipment permits use of electrodes unencumbered by shielding, allowing easy access for material handling. However, there are disadvantages: it is more expensive and much larger than unstabilized units; adjustment and maintenance are less simple; in many instances, the allowable frequencies do not fit the needs of the process. Not much stabilized equipment is used in the United States; however, it is used widely in Europe.

For operation outside the assigned bands, radiation is limited by shielding or screening the generating equipment and the electrodes. The shielding may be an integral part of the equipment, or it may be a screened room in which one or more unshielded units are operated. The self-shielded equipment has access doors or hoods for loading materials; for continuous processes, materials are fed through radiation-limiting tunnels to the electrodes. The self-shielding devices reduce accessibility for materials handling, but do permit operation on the most suitable frequency. The least expensive equipment, and the least restrictive of materials handling at the electrodes, is the unshielded equipment in a screened room. Most plastics-welding equipment (Figs. 22 and 23) is in this class, some plants having a hundred or more units in a screened room; the room entrance may restrict the flow of materials somewhat. Compliance with the Federal Communications Commission's rules must be formally certified; it is usually done by the equipment manufacturer at his own plant for stabilized or self-shielded equipment. Unshielded equipment must be measured for compliance, and rechecked periodically, at its operating location; measurements may be performed and formally certified by an independent consulting engineer, by the equipment manufacturer, or by the user.

Improperly operating microwave equipment may emit concentrated energy beams, exposure to which must be avoided. With properly designed and operating equipment, however, there is no industrial health hazard. The generator is designed so that opening doors or taking off panels will automatically turn off any dangerous voltages. The work applicators are generally guarded to prevent the operators from coming in contact with them. In properly designed equipment there are no lethal voltages on the work applicators that an operator might accidentally touch. Touching an electrode would cause a burn, possibly a severe one, not any worse, but possibly deeper, than that caused by touching any thermally hot object.

Bibliography

1. *Report on Priority of Static Induced Current,* by Committee Appointed at the Tenth Annual Meeting of the American Electro-Therapeutic Association, Sept. 27, 1900.
2. E. M. Guyer, *Electronics* **18** (6), 92–96 (1945).
3. D. Ward and R. C. Anderson, *Wood Working Digest,* Aug. and Sept. 1964 (Hitchcock Publishing Co., Wheaton, Ill.).
4. J. C. Harper and C. O. Chichester, *Amer. Vacuum Soc. Trans. Tenth Natl. Vacuum Symp., Boston, Mass., Oct. 16–19, 1963.*
5. D. H. Rest, Ref. 4.
6. M. R. Jeppson, *Food Eng.* **36,** 49–52 (Nov. 1964).
7. J. F. Trembley, *SPE J.* **15,** 543–545 (July 1959).
8. "HF Heats Large Bulk Uniformly and Quickly," *Electrified Industry,* July 1964.
9. H. P. Zade, *Heatsealing and High-Frequency Welding of Plastics,* Temple Press Books Limited, London, 1959, p. 79.
10. A. R. von Hippel, *Dielectric Materials and Applications,* The Technology Press of M.I.T. and John Wiley & Sons, Inc., New York, 1954, pp. 18–40.
11. F. Clark, *Insulating Materials for Design and Engineering Practice,* Interscience Publishers, a division of John Wiley & Sons, Inc., New York, 1962.
12. Tables of Dielectric Materials, pp. 301–370 of Ref. 9.
13. *Modern Plastics Encyclopedia Issue for 1965,* Vol. 42, No. 1A, McGraw-Hill Book Co., Inc., New York, 1964.
14. J. E. Goodemote, S. D. Marcey, and E. J. Morrisey, "How to Increase Your Output of Small Epoxy Casting," *Plastics Technol.* **10,** 60–61 (Sept. 1964).
15. C. W. Heckroth and J. F. O'Brien, *Pulp & Paper* **38** (24), 23–25 (July 27, 1964).
16. J. F. O'Brien, *Pulp & Paper,* **38** (38), 27–28 (Nov. 2, 1964).
17. R. Blau, M. Powell, and J. E. Gerling, "Results of 2450 MC Microwave Treatment in Potato Chip Finishing," *Paper, 28th Ann. Conf. Potato Chip Inst. Intern., New York, Feb. 2, 1965.*
18. *Chem. Eng. News* **41** (17), 50 (1963).
19. *Ind. Electronic* p. 411 (May 1963).

General References

G. H. Brown, C. N. Hoyler, and R. A. Bierwirth, *Theory and Applications of Radio-Frequency Heating,* D. Van Nostrand Co., Inc., New York, 1947.
    Contains thorough theoretical and practical analyses of typical problems in wood-gluing, sealing, shaped electrodes, dehydration, and food treatment.

H. P. Zade, *Heatsealing and High-Frequency Welding of Plastics*, Temple Press Books Limited, London, 1959.
> Complete and concise coverage of dielectric heat-sealing; also has excellent short sections on theory of dielectric heating and on generator design; has an extensive bibliography, including U.S. and foreign patents.

L. L. Langton, *Radio-frequency Heating Equipment*, Pitman Publishing Corp., New York, 1949.

R. W. Peterson, *Dielectric Heating as Applied to the Woodworking Industries*, Forest Service Bulletin, Ottawa, 1954.

J. Pound, *Practical RF Heating for the Wood Industry*, Heywood & Company Ltd., London, 1957.

P. H. Graham, "Electronic Gluing," articles in *Wood Working Digest* (a monthly publication), Hitchcock Publishing Co., Wheaton, Ill., 1952 and 1953.
> Extensive detailed coverage of wood gluing and forming applications.

E. C. Stanley, *High Frequency Plastic Sheet Welding*, Radyne Ltd., Wokingham, Berks, England, 1961.
> Much practical information about a large number of applications in plastic welding.

Industrial Applications Laboratory, *Bulletin 1*, Eitel-McCullough, Inc., San Carlos, Cal., Dec. 1964.
> An annotated bibliography of selected articles dealing with various aspects of the use of microwave energy for industrial purposes.

RF and Microwave Dielectric Heating System Design Parameters, Industrial Applications Laboratory, *Bulletin 2*, Eitel-McCullough, San Carlos, Cal., March 1965.
> Short theoretical analysis and sample calculations.

A. R. Von Hippel, *Dielectric Materials and Applications*, The Technology Press of M.I.T. and John Wiley & Sons, Inc., New York, 1954.
> Has a theoretical analysis of the physics of the heating phenomenon, and a complete and extensive tabulation of dielectric properties of materials.

J. F. Trembley, "Preheating for Thermoset Preforms and Other Dielectric Materials," paper available from W. T. LaRose & Associates, Inc., Troy, New York, Oct. 1963.

Milton Rothstein
The Thermatron Company
Industrial Electronics Division
Willcox & Gibbs Sewing Machine Company

ULTRASONIC FABRICATION

The first practical uses of ultrasonic energy were in World Wars I and II, in the underwater detection of submarines. The first application to plastics came in the 1950s when films were joined ultrasonically. But it was not until 1963, when ultrasonic energy was first applied to rigid thermoplastics, that the full potential of ultrasonic assembly operations began to be appreciated. Since that time, advances in equipment, joint design, and techniques have made ultrasonic assembly an important method of joining thermoplastics in the automotive, electronics, furniture, toy, and appliance industries.

There are five methods of ultrasonic assembly: plastics welding of injection-molded parts, metal inserting, staking, spot welding, and sewing. In each application, ultrasonic vibrations above the audible range generate localized heat by vibrating one surface against another. Sufficient frictional heat is released, usually within a fraction of a second, to cause most thermoplastic materials to melt, flow, and fuse. Ultrasonic assembly is cleaner, faster, and more economical than conventional bonding methods. It eliminates the application of heat, solvents, or adhesives, and does not require any curing time. Therefore, it permits higher production rates. See also ADHESION AND BONDING; SOUND ABSORPTION.

Equipment

The first step in producing ultrasonic energy is to change 60-Hz electric current to 20 kHz. This high-frequency current is then fed into a converter which changes the electrical energy into 20-kHz vibratory energy. The vibratory energy is transferred to the mechanical impedance transformer, which in industry is commonly called a "horn" because it is a tuned resonant section. Ultrasonic assembly equipment also includes a stand and a timer, or programmer.

The Power Supply. The purpose of the power supply is to modify the line current to appropriate frequency levels. Existing systems have power outputs at various levels up to 35,500 inch-pounds per second at the tip of the horn (see the section on Converters for a discussion of rating systems). All of the power-supply functions operate from a 117/220 volts AC, 50/60 Hz current, depending on the power required for the equipment and the local power source. Power supplies vary in output and may have vacuum-tube or, more often, solid-state circuitry. Manual controls cover frequency fine tuning and output power levels, while appropriate circuitry provides automatic frequency control.

The Converter. The converter or transducer transforms 20 kHz electrical energy, as received from the power supply, into mechanical energy vibrating at 20 kHz per second. There are two methods of conversion: electrostrictive and magnetostrictive. *Electrostrictive conversion* uses a piezoelectric element of lead zirconate titanate, which expands and contracts when excited electrically. Piezoelectric transducers operate with efficiencies in excess of 90%, depending upon the design.

In *magnetostrictive conversion*, the transducer core changes its length under the influence of an alternating magnetic flux field. The core is usually made of nickel alloy. Exciting coils are wound around the core to produce the magnetic flux. The magnetostrictive transducer is limited to less than 50% efficiency, owing to resistance and hysteresis losses.

Ultrasonic energy is rated either in watt output delivered by the power supply or in inch-pounds per second delivered at the tip of the horn. Because of differences in a transducer's conversion efficiency, wattage ratings may not be applied with the same meaning to all systems. Therefore, values in inch-pounds per second measured at the output end of the horn more nearly reflect the capabilities of the equipment, since all losses in the converter have been accounted for.

The Horn. After the electrical energy has been converted into mechanical energy, it is transmitted through the horn, or mechanical impedance transformer. The horn's function is to achieve the proper amplitude and the required energy transfer for the operation. Horns are, therefore, shaped to fit the part closely. The horn is tuned to resonate at the system's frequency, and is usually one-half wavelength long. Materials used in horn construction must have high strength-to-weight ratios in addition to good acoustical properties. Most horns are made of titanium, although some have been made of Monel alloy, beryllium, or aluminum.

The mass and shape of the horn as well as the physical properties of the material of construction influence the length at which it will resonate at the required frequency. More complex shapes and larger sizes are being made possible because of better understanding of acoustics and more sophisticated test equipment.

The Stand. The stand supports the converter–horn assembly in the proper relationship to the workpiece, and controls operating pressure, contact rate, and triggering. Pneumatic systems, because of their adaptability and ease of adjustment over a broad range, are used most frequently. High production rates are obtainable through semi- and fully automated systems.

In a typical ultrasonic assembly stand, all control functions are incorporated within the unit. Holding, or clamping, pressure is developed by a pneumatic system within the stand and is exerted by the horn to hold the part being ultrasonically fabricated in position. The regulating system used also filters the incoming air, introduces lubrication, and monitors the operating pressure. Sensing of contact with the part and adjustment of the air flow for motion control complete the features of a well-designed stand.

A pneumo-mechanical sensing of contact with the part activates the delivery of ultrasonic energy, which is then timed by a solid-state electronic programmer. If the unit is incorporated into an automatic materials-handling system, limit switches can sense cycle completion and activate the advancement of a new set of components to be assembled.

Fig. 1. Pistol-grip hand tool.

Stands are bench-mounted units and, therefore, not portable. A pistol-grip hand tool, shown in Figure 1, is totally portable and capable of spot welding, staking, and inserting. It has broadened the scope of ultrasonic assembly to include parts of very large size with hard-to-reach joints. The operator controls the pressure of the tool against the plastic manually and the time of ultrasonic-energy flow with a trigger switch in the grip.

The Programmer. The programmer for ultrasonic assembly systems controls several functions, some of which operate simultaneously, others sequentially. A foot- or hand-operated switch activates the pneumatic system to apply the horn to the part. Contact with the part signals the programmer to deliver energy for a preset period of time, adjustable over a range of 0.1–6.0 seconds. After the welding cycle is completed, a holding phase of 0.05–3.0 seconds permits the plastic to cool while pressure on the assembly is maintained. This allows the material to acquire sufficient strength so that the newly formed joint will not be separated. A final function of the programmer is to cause the horn to be retracted from the workpiece so that a new cycle may be started.

Methods

There are five basic methods of ultrasonic assembly: plastics welding, metal insertion, staking, spot welding, and sewing.

Welding. Ultrasonic welding can join injection-molded materials without the use of solvents, heat, or adhesives. The face of the ultrasonic horn is brought into contact with one of two plastic parts being welded. The horn applies a predetermined degree of pressure against the assembly, and mechanical vibrations at ultrasonic frequencies are transmitted from the horn into the plastic for a controlled period of time. Utilizing the acoustical transmitting property of the plastic, the vibrations travel until they meet the joining surfaces and are released in the form of heat.

Joint design is one of the most important requirements for a good ultrasonic weld, and the key to good joint design is the proper use of an "energy director" (see Fig. 2). An energy director is a triangular section incorporated into a joint design that defines and controls the area at which energy is dissipated. The high concentration of energy focused by the energy director results in almost immediate melting and a uniform flow in the joint area.   The size of the energy director is directly related to that of the section to be welded.   The base of the triangular energy director is 20% of the section width, and the height is 10% of the section.   The angle at the peak is 90°, which facilitates modification of an existing mold to incorporate the energy-director design, and also permits satisfactory molding of the desired profile.   This modified joint permits rapid welding and achieves maximum strength.

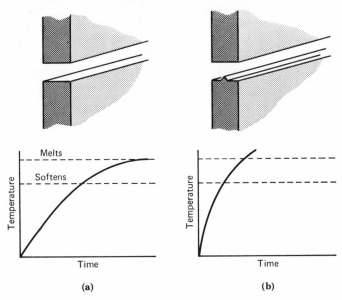

Fig. 2.   Time–temperature relationships in ultrasonic welding of a thermoplastic: (**a**) for a butt joint;   (**b**) for a joint incorporating an "energy director."

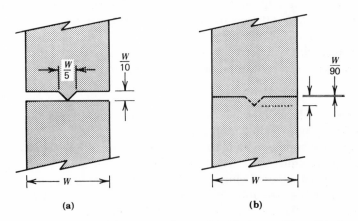

Fig. 3.   Design of a butt joint with "energy director" (**a**) before welding, (**b**) after welding.

Figure 3 shows a simple butt joint modified with an energy director of the desired proportions before and after welding.   The material from the energy director dissipates throughout the joint area, forming a layer of the size indicated.

Figure 4 illustrates a step joint used where a weld bead on the side would be objectionable.   This joint has excellent self-aligning capability and is usually much stronger than a butt joint, since material flows into the clearance necessary for a slip fit and establishes a seal that provides strength in shear as well as in tension.

The part to be welded must be held rigidly. Fixturing nests are provided in many shapes to support the part and locate it under the horn. The nest is usually machined of aluminum or steel. Nests for oddly shaped parts may be molded in epoxy resin or a

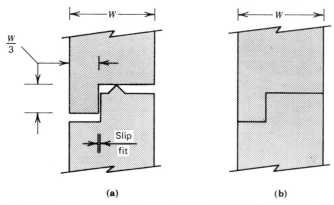

(a)   **(b)**

Fig. 4. Design of a step joint for ultrasonic assembly (**a**) before welding, (**b**) after welding.

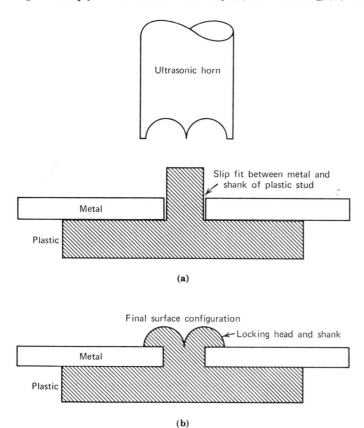

Fig. 5. Staking with ultrasonics (**a**) before contact of horn with plastic, (**b**) after completion of the operation.

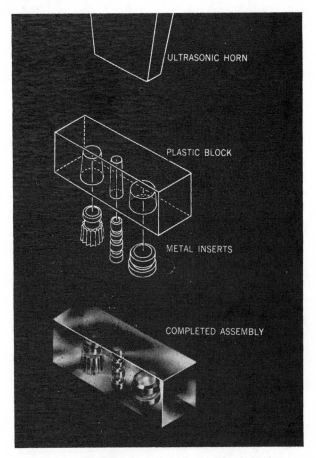

Fig. 6. Ultrasonic inserting of metal into plastic.

similar material (see TOOLING WITH PLASTICS). The nest provides accurate placement and support of the part under the horn. Thin, resilient liners are sometimes used in the nest. This cushioning improves energy flow and reduces marring of the nested part.

The success of welding and staking of plastic parts or inserting metal into plastics depends upon the proper amplitude of the horn tip. Since it may be impossible to design the correct amplitude into the horn initially because of its shape, booster horns may be necessary to increase or decrease the amplitude and produce, thereby, the proper degree of melting or flow. The nature of the plastic, its shape, and the type of work to be performed all determine the optimum horn amplitude.

Staking. Most ultrasonic staking applications involve the assembly of metal and plastic. As with ultrasonic welding, localized heat is created through the application of high-frequency vibrations. In staking, minimal pressure is used so that the energy will be released where the horn comes in contact with the plastic. The specially contoured tip in the horn meets the stud and reshapes it (see Fig. 5). Since frequency, pressure, and time are always the same during each cycle, frictional heat generated between the horn and stud is consistent and much faster than heat applied with a hot iron. The plastic is then permitted to cool for a fraction of a second before horn pressure is released.

Inserting. In many cases ultrasonic inserting can replace the conventional and costly process of insert molding. A hole, slightly smaller in diameter than the insert it is to receive, guides the insert into the plastic while ultrasonic energy and pressure are being applied by the horn. The heat created by the vibration of the metal insert is sufficient to melt a thin film of plastic around the hole momentarily, permitting the inserts to be driven into place (see Fig. 6).

The time required is usually less than one second, but during this brief contact the plastic reshapes itself around knurls, flutes, undercuts, or threads to encapsulate the inserts (see also EMBEDDING). The horn can be contoured to match the contact surface. Parts normally remain relatively cool because heat is generated only at the plastic–metal interface. The size of the metal components and their number determine the exposure time; many small inserts require as little as 0.1 second.

Fig. 7. Ultrasonic spot welding of two sheets of clear thermoplastic.

Ultrasonic insertion is advantageous because it can be carried out as a postmolding operation, and tolerances of the inserts are not critical.

Spot Welding. The new technique of spot welding thermoplastics with ultrasonic energy requires a specially shaped tool. Vibrating ultrasonically, the point of the tip penetrates the first sheet and displaces molten plastic. This material is shaped by a radial cavity in the tip and forms a neat, raised weld on the surface. Simultaneously, energy is driven through the contacted sheet and released at the interface. This produces frictional heat between the sheets to be welded. The molten plastic displaced when the tip comes in contact with the second sheet flows between the two heated sheets and forms a permanent bond. Spot welds may be made with a pistol-grip hand tool or a pneumatic tool, both of which are portable, or with a bench-model ultrasonic assembly equipment (see Fig. 7).

Spot welding requires out-of-phase vibrations between the horn and plastic surfaces. Light initial pressure is used for the out-of-phase activity at the contact area. In manual operation, proper depth is reached when the rate of penetration suddenly increases; this indicates that the tip has passed through the heated interface area. Energy is immediately turned off while pressure is retained for a fraction of a second.

A compression collar on the ultrasonic assembly equipment assures uniform welds. An adjustable stop is incorporated within the design which also controls depth of penetration.

Sewing. In ultrasonic sewing as in other methods of ultrasonic joining, ultrasonic energy is generated by a solid-state power supply, converted to mechanical vibrations, and applied to a tool called a horn. Thermoplastic materials are fed between the tip of the horn and a specially designed anvil called a stitching wheel. The vibrating horn creates frictional heat within the fabric where the fibers are compressed between the horn and wheel, causing the material to melt and fuse in the pattern of the wheel.

Since the bond is dependent upon melting, ultrasonic sewing is limited to synthetic thermoplastic materials such as nylons, polyesters, polypropylenes, modified acrylics, some vinyls, and polyurethans. Blended materials with up to 35% natural fiber content can also be sewn ultrasonically. Ultrasonics can sew knitted, woven, nonwoven, or film materials.

Ultrasonic sewing equipment can be used to seam, hem, tack, baste, pleat, slit, and buttonhole for such applications as bagging, bandages, blankets, clothing, disposables, draperies, film, filters, packaging, sails, strapping, upholstery, and wi ndo shades. Numerous stitching designs are available; see Figure 8.

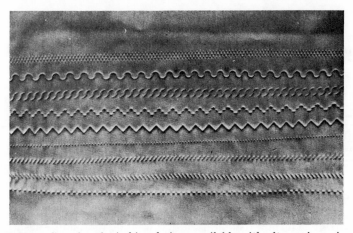

Fig. 8. Samples of stitching designs available with ultrasonic sewing.

Staking and Welding Characteristics of Plastics

Most commonly used injection-molded materials can be ultrasonically welded. Weldability depends on melting temperature, modulus of elasticity, impact resistance, coefficient of friction, and thermal conductivity. Generally, the more rigid the plastic, the easier it is to weld. Low-modulus materials, such as polypropylene and polyethylene, can often be welded provided the horn can be positioned close to the joint area. In staking, the opposite is usually true. The softer the plastic, the easier it is to stake. However, good results can be achieved with most plastics when the right combination of amplitude and force is used.

Different plastics must be chemically compatible and have similar melt temperatures to be weldable to each other. For example, polyethylene cannot be welded to polystyrene because the polymers are not chemically compatible and the melt tem-

Table 1 Suitability of Plastics in Ultrasonic Fabrication[a]

Plastics	Staking and inserting	Welding[b]		Remarks
		Near-field	Far-field	
polystyrene				
unfilled	E	E	E	produces strong, smooth joints
rubber-modified	E	E	G–P	welding characteristics depend on degree of impact resistance
glass-filled	E	E	E	weldable with filler content up to 30%
SAN[c]	E	E	E	particularly good as glass-filled compounds
ABS[d]	E	E	G	can be bonded to other polymers, such as SAN, polystyrene, and acrylics
polycarbonates	E	E	E	high melting temperature requires high energy
nylons	E	G	F	levels; oven-dried or "as-molded" parts
polysulfones	E	G	G–F	perform best, owing to hygroscopic nature of the materials
acetals	E	G	G	require high energy and long ultrasonic exposure because of low coefficient of friction
acrylic	E	E	G	weldable to ABS and SAN; applications include dials, radio cases, and meter housings; in sheet form, joints must be machined
poly(phenylene oxide)	G	G	G–F	high melting temperatures require high energy
Noryl[e]	E	G	E–G	levels
phenoxy	E	G	G–F	
polypropylene	E	G–P	F–P	horn design for welding is particularly critical;
polyethylene	E	G–P	F–P	filled compounds usually better, but need individual testing
cellulose esters	G–F	P	P	weldability varies with formulation and configuration; usually perform well in staking and inserting; decomposition of some formulations may occur
vinyl chloride polymers	E–F	F–P	F–P	

[a] Key: E = excellent; G = good; F = fair; P = poor.

[b] Near-field welding refers to joint $\frac{1}{4}$ inch or less from area of horn contact; far-field welding to joint more than $\frac{1}{4}$ inch from contact area.

[c] Styrene–acrylonitrile copolymers.

[d] Acrylonitrile–butadiene–styrene terpolymer.

[e] Polystyrene-modified poly(phenylene oxide), General Electric Co.

peratures are greatly different; the polyethylene would be completely melted before the polystyrene would have reached its melt temperature.

Additives or fillers can detract or add to weldability. Glass fillers tend to increase the stiffness of the plastic, thereby increasing its acoustical response and weldability. Colorants and lubricants occasionally alter such physical properties as the coefficient of friction and melting point, and may reduce weldability. Asbestos- and talc-filled materials may weld poorly because the filler absorbs the molten plastic during the welding process. A summary of the suitability of various plastics for ultrasonic fabrication is given in Table 1.

Bibliography

J. R. Frederick, *Ultrasonic Engineering*, John Wiley & Sons, Inc., New York, 1965.

Jeffrey R. Sherry
Branson Sonic Power Company

ABHERENTS

Abherents or release agents (parting agents) are defined as solid or liquid films that reduce or prevent adhesion between two surfaces. Industrial fields in which abherents have attained great importance include metal casting and processing, food preparation and packaging, rubber and polymer processing, paper coating, the production of pressure-sensitive tapes, and glass fabrication. A number of factors influence adhesion of two materials to each other. The most important ones are penetration, chemical reaction and compatibility, low surface tension, surface configuration, and polarity differences between the two materials. Two solid surfaces generally do not adhere to each other because wetting does not take place and neither does penetration of one into the other. The only exception occurs when one of the surfaces is "tacky" or when chemical reaction takes place between the two surfaces at the interface. Frequently high static charges can also lead to adhesion. Therefore, the use of abherents becomes of technical importance when a solid and a liquid, or even more so, when a solid and a paste or dough form an interface and adhere to each other. For many centuries adhesion of a highly viscous material (paste, dough) has been a disturbing factor in the home in baking and cooking. Abherents in the form of fats, oils, or solids like flour have been used in order to prevent the sticking of dough to wooden kneading boards or to various metal baking dishes. With the greater industrial use of polymeric materials, both natural and synthetic, and also with the industrialization of baking, the commercial use of abherents has become widespread. As a matter of fact, some industries which are of great importance today could not have developed without the availability of modern abherents. As an example let us mention the pressure-sensitive tapes, which could not be unwound if the tape backing were not coated with a release agent.

Properties Required of an Abherent. Since many of the factors causing adhesion are of a chemical nature, one of the first requirements of a good abherent is complete chemical inertness toward the two materials whose adhesion is to be prevented. Adhesion is often due to opposing polarities of the surfaces, therefore the polarity relation of the abherent and of one or both of the surfaces in contact with it have to be taken into consideration in the choice of an abherent. Besides these two factors one physical property is of great importance, namely, good spreading ability (low surface tension), so that the abherent will form a continuous film between the surfaces and in this way exclude any contact between the two materials. An extremely important factor in this action of abherents is temperature dependence. This dependence is so pronounced that a material which is an abherent at, for example, room temperature may act as an adhesive at elevated temperatures. A good example of this type of action is given by polyethylene film, which is used as an abherent layer to prevent the adhesion of uncured rubber slabs or preforms to each other. This same polyethylene film, above its softening point at approximately 100°C, will act as an excellent hot-melt adhesive. Another factor to be considered is volatility. Water would fulfill many of the requirements of an abherent, but because of its vaporization and the increased water absorption of many materials at elevated temperatures, water in most cases is not suited as an abherent. With knowledge of all these factors, we can now narrow down considerably the choice of an abherent. It has to (*1*) be chemically inert toward the two adhering materials, (*2*) have good spreading properties for one or both of the surfaces, (*3*) have low volatility at the temperature at which it is to be used.

In many cases adhesion is caused primarily by the physical configuration of the solid surface, namely, by its high porosity or roughness, which will tend to anchor a viscous material into the surface. In this case a powdered solid material will serve well as an abherent by filling the pores or smoothing out the roughness of a surface. Another example of the usefulness of solid abherents is adhesion caused by the moisture content or surface wetness of a dough; in this case, the solid abherent serves to dry the surface and in this way prevents adhesion.

It is obvious that a liquid, to be effective as an abherent between two materials, will have to spread so as to form a continuous film between the contacting surfaces. Therefore, an additional requirement for a good abherent is a low to medium viscosity at the temperature of application. There is an exception to this requirement, namely, where polymer melt flow is induced by high pressures. In this case an abherent of higher viscosity is more desirable so that it is not pushed aside and displaced by the polymer melt.

Methods for Applying Abherents. As mentioned previously, one of the most important systems in which abherents are used is at the interface of a solid and a dough or paste (polymer melt).

Three basic methods for the application of an abherent in this system are (*a*) spraying, brushing, or dusting a powdered solid or liquid abherent to the solid surface (*b*) producing a permanent abherent by baking an abherent polymeric surface to the solid surface or (*c*) incorporating the abherent in the polymer. In this latter case an abherent has to be chosen that is partly compatible with the polymer, at least at room temperature, and that will exude to some extent at the melt temperature in order to be available at the interface where adhesion has to be prevented. Many polymers will show a slight amount of tack even in the solid stage. This property has caused considerable difficulties in the processing of some polymeric materials. A good example is given by polyethylene films, which tend to adhere to each other because of static electricity and also because of cold flow. Polyvinyl chloride films show the same behavior. Therefore, the incorporation of abherents into these polymers before they are processed into films is standard practice. In the case of polyvinyl chloride films, these additives are called antiblocking agents, whereas in the polyethylene field they have become known as slip agents, although in either case they really serve the same functions.

Industrial Fields Using Abherents

Metal Processing. Two metal-forming processes depend heavily on the use of abherents for their proper functioning. One is *die casting*, in which a rather low-melting alloy is molded under high pressure by transferring the melt to a steel mold and cooling under hydraulic pressure. Abherents are generally sprayed onto the mold surface in order to prevent the casting from sticking to the mold. Because of the high temperatures involved (above 800°F), silicones are the preferred abherents. The other metal-forming method which depends heavily on the use of abherents is the so-called *shell-molding* process. In this process a metal pattern or master is used, around which is poured a sand and resin combination which eventually forms the mold for casting the metal part. The metal pattern or master has to be re-used many times to form new shells, and fast and easy release of the shell from the pattern is therefore of great importance. Here again silicones have made an important place for themselves. A new pattern is usually coated with a curable silicone paste and, in continuous use, is usually recoated by spraying with a silicone emulsion after each cycle.

Silicone greases are also used to coat oven conveyors, dollies, carts, and other handling equipment that must operate under high-temperature conditions. The use of release agents in shell molding has led to tremendous improvements in this operation since the sticking of sand to the pattern after curing has been very detrimental and has slowed down the operation, necessitating cleaning of the patterns after each use.

Release agents are also used in the metal industry for the molding or casting of ingots from zinc, brass, bronze, lead, and precious metals. The release agents, primarily silicones, are applied to the mold in order to give easier release of the castings and to avoid build ups in the molds.

Food Industry. As mentioned above, release agents have been used for centuries in the baking and frying of various foods. In this case the choice of an abherent was, of course, limited to edible materials. In very recent years however, the use of inert coatings for baking or frying pans has been more and more accepted. Frying pans for the home are now available with a polytetrafluoroethylene coating, which is permanent if treated carefully and not injured by handling with metal tools. In industrial baking, where vegetable oils and fats have been used for many centuries, a tremendous improvement has been achieved by coating baking pans with a silicone varnish which gives the pan long-term abherent properties.

Rubber Processing. See Elastomers, synthetic; Rubber, natural. Abherents have attained extreme importance in the processing of rubber. Generally, both natural and synthetic rubbers have excellent adhesive qualities in the uncured stage, yet these rubbers have to be processed and shaped before they can be cured. There are a great variety of processing steps involved in the manufacture of rubber goods. Generally the procedure may be broken down as follows: First, natural, synthetic, or mixtures of both rubbers are blended with various compounding ingredients, such as fillers, accelerators, vulcanizing agents, pigments, and other ingredients that go into specific rubber compounds. This is usually accomplished in an internal mixer of the Banbury type. Then, this mixed batch of rubber is generally sheeted through a rubber mill to put it into the form of a sheet or slab, which may then be calendered or molded into thinner sheets, extruded into tubing, or formed into other preforms for molded rubber goods. During these processing steps that precede vulcanization, the rubber compounds would tend to stick to metal surfaces and even more so to each other. In order to prevent this sticking, coated papers and coated cloths have been used for many years to separate the individual layers of uncured rubber. Also, ever since rubber processing has become an industrial art, various abherents have been used to dust or spray onto the molds. Abherents have also been incorporated into the compounds to reduce tackiness.   All of the various types of abherents have been and are still being used in the rubber industry.   Of primary importance are the various metal stearates, stearic acid, oleic acid, microcrystalline and paraffin waxes and other synthetic waxes, like the stearamides, alkyl *N*-substituted stearamides, and bis-stearamides (derived from diamines, eg, ethylenediamine), and ester waxes, such as montan wax.   In rubber processing, as in many other fields where abherents have been employed, the advent of the silicones has brought about major changes.   Previously, molds for molding rubber goods or for curing rubber sheeting had been dusted after each cycle with zinc stearate, calcium stearate, and similar abherents.   These have mostly been replaced by silicones in the form of water dispersions, which are sprayed onto the molds in dilutions of 1% or below.   Whereas previous lubricants had led to build up of crusts in the molds and had necessitated frequent cleaning, the silicones have eliminated this shortcoming completely.   One other use of abherents in the rubber industry is the dusting of finished unfilled rubber goods which would be tacky without a surface application of abherent.   In this case solid materials have been found to serve best; talcum, mica, finely dispersed silicates, or metallic stearates have proved advantageous.

Stearic acid, one to five parts per hundred of rubber, usually forms part of any rubber formulation.   The function of this fatty acid is not completely clear.   It is assumed that the stearic acid contained in both natural and synthetic rubbers, in addition to that which is added to the compound, reacts with the zinc oxide present in every formulation during vulcanization; furthermore, that the zinc stearate thus formed, in turn, reacts with the accelerator, enabling the latter to promote the vulcanization.   Without any doubt the addition of fatty acid also produces an abherent effect in this case.

Polymer Processing.   The use of abherents is of the utmost importance in polymer processing.   Some polymers have particularly great adhesive properties at or about their melting points and are, therefore, in greater need of abherents than others.   Examples of these highly adhesive melts are polystyrene, some polyolefins, other hydrocarbon resins, methyl methacrylate, and, to some extent, polyamides. But even polymers with lower adhesive forces require the use of release agents in most cases.   The above-mentioned polymers, which are all thermoplastic in nature, are not anywhere near as high in their adhesive properties as some thermosetting resins that have to be cured in contact with the mold surface.   Examples of these latter are the polyesters, polyurethanes, and polyepoxides.

The use of abherents in polymer processing extends to all possible ways of application.   Most polymeric compounds made by the raw material manufacturers already contain abherents.   Furthermore, the processor who adds pigments and other ingredients to the raw resins also adds abherents at that stage.   In many cases the fabricator who molds or extrudes thermoplastics will tumble some release agent or abherent onto the molding powders or granules before melting them in the processing equipment.   Even beyond this preparatory addition of abherents, an application of abherents to the mold surfaces or metal surfaces of other processing equipment is frequently necessary.   From the number of ways in which abherents are added to polymers, the importance of these products in various stages of polymer processing becomes apparent.

In order to illustrate the application of abherents, a few specific examples will be cited.   In the injection molding of polystyrene, which is the greatest volume injection-molding material in use, abherents are added when the raw material is processed. Pure polystyrene as it comes out of the polymerization process needs the addition of abherents to promote the conversion of the raw polymer into molding pellets or granules. This conversion, which is usually accomplished through an extruder, is greatly facilitated by the addition of small amounts of abherents (metallic stearates, stearic acid). The increase in throughput through the extruder, when materials containing abherents are compared to those not containing any, can be as much as 20%. These molding pellets or granules, after they have been cooled, are again tumbled with a parting agent, which more or less adheres to the surfaces of the individual pellets. In this case a great variety of abherents can be applied. Zinc stearate, stearic acid, and stearamide are among the products most frequently used, but room-temperature liquids such as butyl stearate, are also employed in many cases.

In the injection-molding process unmelted molding pellets are pushed by a hydraulic ram into a cylinder where they are gradually heated and melted. It is the function of the abherent to reduce the friction of the granules against each other and also against the metal surfaces of the cylinder in order to reduce the pressure needed to convey them forward into the melting zone. This polymer melt is then injected by the ram into a cold form or mold. As soon as the polymer melt has sufficiently solidified, this shaped item can be removed from the form and the added abherent again helps tremendously in releasing the item from the metal surface. Nevertheless, in many cases additional abherents or release agents have to be applied to the mold surface in order to further facilitate the release of the molded polymeric article from the metal mold. It can easily be seen that the function of abherents in facilitating the molding of a polymer will also reduce the temperature to which the material has to be heated for it to flow properly into the mold. Since high temperatures are detrimental to the stability of most polymers, the abherents therefore also have a protective function, and result in articles of greater strength and quality. It should be noted that silicones are not used as abherent additives to polymers. The reason is that silicones are so powerful as abherents and so inert and noncompatible that their addition to polymers would counteract the cohesive forces and destroy the homogeneity of the material.

Silicones make very effective abherents when applied to the mold in the injection or compression molding of plastic materials but they have to be used with great caution, because molded polymeric items often have to be adhered to other surfaces, or lacquers have to be applied to them for decorative purposes. Silicones remaining on the surface of these molded pieces would seriously interfere with the above-mentioned aftertreatments.

In the processing of rigid and flexible polyvinyl chloride, abherents are also of great importance although polyvinyl chloride does not have as strongly adhesive a character in the melted state as polystyrene. In the manufacture of films from polyvinyl chloride, the compounds are generally blended in a cold or somewhat heated blender. They are then melted and fused on a heated rubber mill or in an internal mixer. This melt is then fed into a calender consisting of three or four heated rolls through which the polyvinyl chloride compound is squeezed down to the thickness desired in the finished film. For calendering, abherents have to be skillfully compounded into any polyvinyl chloride material, in order to make the melt follow the proper rolls and to make it release from the ones that should not be followed. Frequently the use of one abherent is not sufficient, and a combination of two or more have to be applied. An additional function of the abherent lies in the prevention of blocking or sticking of the film to the next layer in the roll of the finished film. Whereas small amounts of stearic acid provide excellent release from metal surfaces, other abherents, like calcium stearate, lead stearate, and stearyl amides are added to provide a combination of release from metal and antiblocking properties. These latter abherents will also greatly influence the surface appearance of the calendered film, providing such desirable properties as gloss and smoothness.

In the extrusion of polyvinyl chloride, a process that consists in melting a granulated or powdered polyvinyl chloride compound between a cooled rotating screw and a heated cylinder and in then pushing this homogenized heated melt through a shaping orifice, the addition of abherents will help release the compound from the rotating screw and in this fashion greatly increase the speed with which the compound is conveyed through the extruder cylinder. The abherents will also reduce the possibility of thermal breakdown, which is critical with polyvinyl chloride. Since the shaping orifice or die is usually the hottest part of the extruder, easier release from this hot and normally glossy surface is provided by coating the die surface with a silicone paste or grease prior to extrusion. This serves to fill in the pores of the metal and to prevent carbonization of the polymer melt as a result of prolonged contact with the hot metal. Coating of the die surface with abherent, however, is no substitute for the addition of various abherents to the compound itself.

In the casting of thermosetting plastic materials, such as polyesters and poly-epoxides, it is most essential to provide easy release from either the male mold over which such a compound is cast or the female mold and flexible bag in which it is molded. Matched metal molds are also sometimes used to shape these compounds. Polyepoxides are known to be outstanding adhesives. Where release from a mold surface has to be provided, the use of abherents in one fashion or another is an absolute necessity. It is interesting to note that, in this field of casting thermosetting materials, the use of polymeric films as abherents, in the form of extruded or cast films, has been found advantageous. Polyamide or polyvinyl alcohol films have proved highly successful in this application. In addition to these films, sprayed-on silicones or waxes are also frequently employed. Although many more such examples could be cited, it is hoped that the above will show the need for abherents and their widespread use in the field of polymer processing.

Paper Coating and Pressure-Sensitive Tapes. Adhesive coated papers, cloth, and plastic films have attained great industrial importance during the last twenty years. All these materials, after being coated with adhesive and dried, have to be wound up in rolls. Abherents are needed to prevent the rolled-up adhesive coating from sticking to the backing material. These applications make very severe demands on abherents, and considerable work has been done in this field in order to develop the proper materials.

Here again, as in so many applications, the silicones have done an outstanding job. Silicone-coated papers have attained great commercial importance and are being manufactured by a large number of specialty paper manufacturers. These papers, which are coated with the silicone abherents on either one or both sides, are used as interleafing papers for various sticky substances like uncured rubber, whereas the papers coated on one side only are used primarily as the base material for pressure-sensitive and other adhesive paper tapes. Before the introduction and acceptance of silicone coatings, wax–polyethylene coatings, straight wax coatings, and zinc stearate or talcum dusting had been used to a great extent.

An application that combines both paper coating and polymer processing provides another outlet for abherents. Polyvinyl chloride plastisols, which are dispersions of finely powdered polymer in liquid plasticizers, are frequently cast into sheets or films that are then fused into a polymeric compound. Originally, these films were cast onto stainless-steel continuous belts, an operation that entailed a considerable initial investment as well as replacement cost. These films can now be cast and fused on abherent-treated papers, and the films are peeled off the papers after fusion.

Abherent-treated papers are also used widely in the general packaging of sticky materials. These include chemicals, foods, and many others. Practically all classes of abherents are used in the packaging field.

Glass Industry. The uses of abherents in the glass industry and, more particularly, in the molding of glass, are similar in importance and application to their use in polymer processing. In glass molding, the various ingredients are melted together in large furnaces. From the melting furnace the glass melt is delivered through feeding channels, where it is cut by shears into gobs of desired size. These gobs are delivered by conveying devices to the molding stations, first to a blank mold and then to a finish mold. Subsequently, these molded bottles or other items, still very hot, are conveyed through a heated annealing furnace and are cooled slowly to release internal strains.

All metal parts coming in contact with the high-temperature glass melt have to be treated with abherents to prevent sticking of the melt to the metal. Since the glass molding industry dates back almost a hundred years, the first abherents were natural waxes and vegetable and animal oils. Later, mineral oils were used and today, silicones are employed most often. Silicones have been so very valuable in this application because of the high temperatures required in glass processing. The vaporization and carbonization of organic abherents causes fumes and crusting. The use of silicones has therefore brought about a major improvement in the speed, efficiency, and cleanliness of this operation.

Classes of Abherent Materials

WAXES

Both natural and manufactured waxes (see also Petroleum waxes) are finding application as abherents. Natural waxes that have gained importance in this field are paraffin and microcrystalline waxes (both types are petroleum products); waxes of vegetable origin, such as carnauba or candelilla wax; and waxes of animal origin, such as spermaceti.

The petroleum-based waxes have the disadvantage of relatively low melting points, which may produce stickiness near the melting point, rather than release. These waxes also are subject to oxidation at high temperatures. The vegetable waxes, like carnauba wax, are excellent abherents but, being natural products, have the disadvantage of variation in color and price. High price also has been limiting the use of spermaceti wax.

Synthetic or manufactured waxes have attained greater importance for use as abherents. Practically all aliphatic alcohols from the C-10 up have found use as abherents. The same can be said about fatty acids above C-12. The fatty acid having attained the widest use as an abherent is stearic acid, which is available highly purified at a reasonable price. It has a sharply defined melting point and good wetting properties. Since stearic acid also has a very limited compatibility with organic polymers, it has been found an efficient abherent in a great number of applications. As examples, one may mention the internal lubrication produced (by addition to the formulation) in polyvinyl chloride and styrene polymers and copolymers. The use of stearic acid in rubber compounds, where it not only serves as an abherent but performs other functions as well, has been covered previously.

The glyceryl stearates and various glycol stearates comprise another group of synthetic waxes that are employed as abherents in a multitude of applications. Two important members of this family are (*1*) glyceryl monostearate (α-stearin), used as an additive in polyvinyl chloride compounds for the manufacture of sheeting and film, and as an additive in polybutenes (Vistanex), which otherwise have exceptional adhesive properties, and (*2*) diethylene glycol monostearate, used as an incorporated abherent in rubber processing and in polyvinyl chloride compounds and as an abherent coating on paper. Another interesting group in this family of glyceryl fatty acid esters are the hydrogenated oils. Of particular industrial interest is fully hydrogenated castor oil, which is sold under the trade name of Opalwax. Chemically, Opalwax is glyceryl tri(hydroxystearate). This wax has an exceptionally high melting point and is used as an abherent in rubber compounds, coated fabrics, and papers. Probably the most important single abherent in the family of synthetic waxes, known under various trade names (Acrawax C, Advawax 280), is reported to consist primarily of ethylene bis-stearamide. This is an amorphous wax of good color, excellent heat stability and a melting point of 140–143°C, probably the highest of all the waxes. This wax, having such a high melting point, can be obtained and shipped in a very finely pulverized form, which makes it more convenient to apply than most waxes. It also has extremely limited solubility and compatibility and, therefore, can be used in a great number of systems. Acrawax C has been found useful as an abherent in almost all types of polymer processing and, in many cases, acts in the finished polymer film or sheet as an antiblocking agent, or slip agent, preventing adhesion of two layers of film to each other. This wax has also found many uses as a release agent in metal processing.

METAL SALTS OF FATTY ACIDS

The metal salts of fatty acids and primarily of stearic acid (see Driers and metallic soaps) have acquired an important industrial position as abherents. As high-melting solids usually available in a powdered form, these metal salts are applied primarily as dusts, but also in many cases are incorporated into polymeric compounds as internal abherents or release agents. In rare instances water dispersions of these metallic salts have also been applied. Zinc stearate, for example, is available in a water dispersion and has been used in this form in rubber processing. The most important metallic salts used as abherents are calcium stearate, zinc stearate, lead stearate, magnesium stearate, and aluminum stearate. Which metallic salt is chosen for a specific application depends primarily on the polymers and other surfaces involved. Calcium and lead stearate are the dominant abherents in polyvinyl chloride processing, where both also have heat stabilizing effects. Zinc stearate is substituted for lead stearate in specific cases where nontoxicity is a requirement, but it does not have anywhere near the stabilizing effect of lead stearate. Calcium stearate is probably the most effective abherent in polyvinyl chloride.

As an abherent in polystyrene, zinc stearate has found the greatest use. Aluminum stearate is also effective with many polymers. In the field of rubber processing, aluminum and magnesium salts are the preferred abherents. Zinc and aluminum stearates have found widespread use in metal processing, particularly in the drawing of wire, in sheet-metal stamping, and in powder metallurgy. The reasons for the superior functioning of one metallic salt over another in various applications have been determined empirically. Frequently the choice of the right stearate can be based on melting point, insolubility, and particle size to which the stearate can be ground or precipitated.

Metallic salts are used indirectly as abherents where a low-melting wax is the primary abherent and the metallic salt is employed to increase the melting point of the mixture. Thus, blends of aluminum stearate and paraffin wax and also of zinc stearate and stearic acid have been made in order to adjust melting points and physical properties of the primary abherent.

POLYMERIC ABHERENTS

Polyvinyl Alcohol. (See Vinyl compounds.) This polymer, which is completely water-soluble and incompatible with practically all organic polymers, has found a variety of uses as an abherent. It is applied as a coating from a water solution, or in the form of a cast or extruded film. One of the major applications is in the molding or lay-up forming of polyesters and epoxides.

Polyamides (qv). Since polyamides are insoluble in most of the commonly used solvents, they find application only in the form of extruded films. These are used with polyester and epoxide lay-ups and are draped over the metal or plaster mold in order to effect release of the cured formed shape from the pattern.

Polyethylene. (See Olefin polymers.) This polymer is used as an abherent film (in the form of extruded tubular or flat film) in the processing and shipping of uncured rubber, and as a paper laminate in the packaging of sticky materials.

Silicones (qv). Silicones represent the most important class of abherents despite their rather recent origin. The generic name "silicones" is applied to a wide range of compounds consisting of silicon–oxygen chains with carbon-containing side groups, such as methyl, ethyl, or phenyl, attached directly to the silica. The commercially useful silicone abherents are all polymeric in order to obtain high boiling points and, therefore, low volatilities at room temperature, heat resistance, and resistance to oxidation. The higher the molecular weight, the greater the chemical resistance and, of course, also the viscosity.

No statistics are available as yet to show the tremendous importance of silicones (in various forms) as abherents, but it is clear that they have replaced other, conventional, products to a very great extent. Prices of the silicones are in the range of about 10 times the cost of conventional abherents, yet in actual use they are in many cases considerably cheaper, because of the very small amounts needed (often they are used in highly diluted solutions or water dispersions below 1% silicone content), and also because they do not build up on metal surfaces as metal stearates do. As a result, they save considerable time in the cleaning of molds or metal casting forms. In the field of polymer processing the only word of warning that applies to the use of silicones is that their abherent properties are too good. In many cases, where surface treatments in the form of printing inks or adhesives have to be applied later, the silicones will interfere with adhesion.

The silicones are used in three forms: fluids, resins, and greases.

Silicone Fluids. Commercially, the most important group of fluid abherents are based on dimethyl silicone. They can be applied either full strength or in solution, or in a water emulsion. These silicone fluids have almost every property required of an ideal abherent:

1. Excellent heat resistance and stability.
2. Low surface tension.
3. Great chemical inertness, especially toward organic materials and polymers.
4. Colorless appearance and nonstaining qualities.
5. Physiological inertness.

Their outstanding properties, together with the ease with which they can be applied, make the silicones ideal abherents in many instances. In applications such as metal processing and glass molding, where extremely high temperatures are involved, the silicones actually are the only abherents that will allow continuous use with great ease.

Silicone Resins. These are cured by heat and catalysts which usually are metallic salts or organic amines. They cure by a condensation reaction in which, by removal of water, oxygen bridges are formed between chains. The more crosslinking occurs, the harder the resin coatings will be. Crosslinking, of course, is favored by a higher functionality of the basic building block. Whereas the silicone fluids usually possess two carbon-containing groups to each Si in the chain, the base groups for resins usually contain only one carbon group for each silica, the rest being hydroxyls.

Silicone resins are manufactured and sold in solution in organic solvents since the solution increases their shelf life. They can be applied by spraying, dipping, or brushing. They are used as abherents primarily on metal surfaces and have found their most important application in baking pans in commercial bakeries. Here again the importance of the silicone resin coating lies in the extreme heat resistance of the fully cured films, which can be used up to approximately 400°F. A paint formulated from silicone resin with powdered aluminum pigment can attain heat resistance up to 500°F.

Silicone Greases. These are silicone fluids thickened with either lithium octoate or with a silica gel, such as Cab-o-sil. The greases have the advantage of not running off too easily, even at elevated temperatures, and therefore acquiring somewhat more permanence than a silicone fluid would when applied to metal or other surfaces.

Fluorocarbon Polymers. The fluorocarbon polymers (see Resins containing fluorine) also are a relatively new group of materials used in many forms. They can be formed into sheets, rods, and other shapes. The fluorocarbon polymers are also available in water dispersions, but their general use as abherents has been retarded by their very high cost.

They do have the properties required of good abherents, namely, very high chemical inertness and high heat resistance. They are employed as abherents in the form of films and sheets, primarily where permanent gasketing between two sticky surfaces is required. Their use, of course, is very limited since a fluorocarbon film or sheet sells for well above $10.00 per pound. One application of polytetrafluoroethylene, as mentioned previously, is as a coating for metal frying pans for home use. This coating is applied as a dispersion, which is dried and fused at approximately 250°C.

INORGANICS

Inorganics are the oldest abherents known. Owing to their insolubility they are used strictly in the form of powders, which exert abherent properties generally because of their flake-like crystal structure. The most important representatives of this class of abherents are talcum (see Talc) and mica (qv). These are always applied as a fine powder sprayed or dusted onto a surface to prevent adhesion between that surface and metal or similar surfaces. They are frequently blended with metal stearates to improve their abherent action. A major application for these inorganic abherents is the dusting of soft rubber products, ie, tubing, rubber bands, and soft rubber sheets. Another example for the use of talcum as an abherent is the dusting of freshly printed paper or polymeric film in order to be able to wind these materials into rolls before the inks are completely dried, without transferring the inks to the backs of the web in the roll.

Bibliography

1. *Commercial Waxes*, H. Bennett, ed., Chemical Publishing Co., Inc., New York, 1956.
2. *Dow Corning Corp. Bulletins 5-116, 5-115b, 5-111, U-5-100, 8-605*.
3. J. W. Keil, "Silicone Paper Coatings," *Tappi* **41**, No. 6 (June, 1958).
4. J. W. Keil, D. L. Leedy, and L. H. Reinke, "Silicone Release Coatings for Paper," *Paper, Film Foil Converter* (August, 1958).
5. R. N. Meals and F. M. Lewis, *Silicones*, Reinhold Publishing Corp., New York, 1959.

George P. Kovach
Foster Grant Co., Inc.

INDEX